MEDIA AND COMMUNICATIONS -
TECHNOLOGIES, POLICIES AND CHALLENGES

HYPERLINK ANALYSIS OF POLITICAL BLOGS COMMUNICATION PATTERNS

MEDIA AND COMMUNICATIONS - TECHNOLOGIES, POLICIES AND CHALLENGES

Additional books in this series can be found on Nova's website under the Series tab.

Additional E-books in this series can be found on Nova's website under the E-book tab.

MEDIA AND COMMUNICATIONS -
TECHNOLOGIES, POLICIES AND CHALLENGES

HYPERLINK ANALYSIS OF POLITICAL BLOGS COMMUNICATION PATTERNS

KOSTAS ZAFIROPOULOS
AND
VASILIKI VRANA

Nova Science Publishers, Inc.
New York

Copyright © 2011 by Nova Science Publishers, Inc.

All rights reserved. No part of this book may be reproduced, stored in a retrieval system or transmitted in any form or by any means: electronic, electrostatic, magnetic, tape, mechanical photocopying, recording or otherwise without the written permission of the Publisher.

For permission to use material from this book please contact us:
Telephone 631-231-7269; Fax 631-231-8175
Web Site: http://www.novapublishers.com

NOTICE TO THE READER

The Publisher has taken reasonable care in the preparation of this book, but makes no expressed or implied warranty of any kind and assumes no responsibility for any errors or omissions. No liability is assumed for incidental or consequential damages in connection with or arising out of information contained in this book. The Publisher shall not be liable for any special, consequential, or exemplary damages resulting, in whole or in part, from the readers' use of, or reliance upon, this material. Any parts of this book based on government reports are so indicated and copyright is claimed for those parts to the extent applicable to compilations of such works.

Independent verification should be sought for any data, advice or recommendations contained in this book. In addition, no responsibility is assumed by the publisher for any injury and/or damage to persons or property arising from any methods, products, instructions, ideas or otherwise contained in this publication.

This publication is designed to provide accurate and authoritative information with regard to the subject matter covered herein. It is sold with the clear understanding that the Publisher is not engaged in rendering legal or any other professional services. If legal or any other expert assistance is required, the services of a competent person should be sought. FROM A DECLARATION OF PARTICIPANTS JOINTLY ADOPTED BY A COMMITTEE OF THE AMERICAN BAR ASSOCIATION AND A COMMITTEE OF PUBLISHERS.

Additional color graphics may be available in the e-book version of this book.

LIBRARY OF CONGRESS CATALOGING-IN-PUBLICATION DATA

Zafiropoulos, Kostas.
  Hyperlink analysis of political blogs communication patterns / Kostas Zafiropoulos, Vasiliki Vrana.
      p. cm.
  Includes bibliographical references.
  ISBN 978-1-61728-922-4 (softcover)
  1. Communication in politics--Blogs. 2. Blogs--Political aspects. 3. Webometrics. I. Vrana, Vasiliki. II. Title.
  JA85.Z34 2009
  320.0285'6752--dc22
                                                    2010025928

Published by Nova Science Publishers, Inc. † New York

CONTENTS

Preface		vii
Introduction		1
Chapter 1	Web 2.0 Applications	3
Chapter 2	Blogs	11
Chapter 3	Analysis of Blogs Communications Patterns	33
Chapter 4	A Social Networking Theory Approach	37
Chapter 5	Exploring Blogs Communication Patterns	43
Conclusion		53
References		55
Index		67

PREFACE

The book provides a literature review on WEB 2.0 applications, particularly on blogging and political blogging, and an analysis of blogs' interconnectivity. It presents recent advances in the hyperlinks analysis of blogs, and describes hypotheses recently introduced in the literature. As a case study, the book examines the degree of interconnectivity, cohesion and polarisation of Greek political blogs that participate in the discussion and use tags about Greek Parties. The study analyses connectivity patterns of Greek political blogs using Social Networking Analysis of the blogs and explores and identifies focal conversational points. The basic hypothesis supported by the literature is that within polarized political systems bloggers locate groups of central blogs, and link to them according to their popularity, familiarity and affiliation. In this way these central blogs gather the majority of incoming links from blogs. Users of the Internet who wish to be informed quickly, locate the focal points of discussion and, for economy of navigation they read only the posts on these blogs. Bloggers also locate focal point blogs and place their posts along with a link to their own blog. They thus expect that the readers of focal point blogs will also visit their blogs. The research examines whether this hypothesis holds for Greek political blogs. Blogs are recorded and their links are analysed using Social Networking Theory and multivariate statistics. Findings reveal that the Greek political blogosphere is connected, though does not present high density. Regarding the formation of central blog groups, Greek political blogs conform well, though not completely, with the communication pattern supported by the literature. According to social network and political characteristics, central blogs are quite partisan and polarised.

INTRODUCTION

Blogs are low-threshold tools for Internet users to personalise and actualise content and information online and to present their views to a broad audience. They are becoming a major source of information and communication for Internet users and have the power to engage people in collaborative activity, knowledge sharing, reflection and debate. Blogs are also featured in papers regarding their political activism and influence, have affected real world events, serve as a medium and tool for political change and offer forms of communication that allow political actors to bypass established media practice.

This book starts by providing a literature review on the political use of web 2.0, blogging and political blogging. It presents the recent advances in the hyperlink analysis of blogs, and describes the recent hypotheses used for this kind of research. As a case study, the book examines the degree of interconnectivity, cohesion and polarisation of Greek political blogs that participate in the discussion using tags about Greek Parties. Tags are labels of thematic categories of blog posts regarding their subject. In order to cope with the research question, the book records Greek blogs with tags about parties of the Greek Parliament. Blogs with tags about Greek parties ND, PASOK, KKE, SYRIZA, and LAOS are considered.

The analysis chapter studies connectivity patterns of Greek political blogs using a two step approach. First, it uses a Social Networking Analysis of the blogs using hyperlinks. Graph theoretic properties and indexes are used to present connectivity properties of the blogs, in order to explore the fashion of blogs self-organization. The second step regards the exploration and identification of focal conversational points. The basic hypothesis supported by the literature is that within polarized political systems blogs are forming clusters around central blogs, which are considered reliable or ones which have the same affiliation. Users of the Internet who wish to be informed quickly, locate the focal points of discussion and also for economy of navigation, they read only the posts on these blogs. Bloggers also locate focal point blogs and place their posts along with a link to their blog. Thus, they expect that the readers of focal point blogs will also visit their blogs. The research examines whether this hypothesis holds for Greek political blogs. The quantitative analysis initially uses the theory of Social Networking. The blogs and their incoming links along with the association matrices (0/1) of the blogs networks are recorded. Multidimensional Scaling and Hierarchical Cluster Analysis are used to locate clusters of focal point blogs. Several quantitative indexes are calculated to present the degree of interconnectivity, cohesion and polarization of blogs tagging to Greek Political Parties. Regarding the Social Network properties of the blogs, the Greek political blogosphere is connected, though without presenting high density. Regarding the formation of central blog groups, Greek political blogs conform well, though not completely, with the communication pattern supported by the literature. Investigation of the political and ideological affiliation of the central blog clusters, along with the study of the volume of connectivity (through hyperlinks), describes a system, which is partisan and polarized to some extent.

Chapter 1

WEB 2.0 APPLICATIONS

WEB 2.0

The term "Web 2.0" was coined by Tim O'Reilly of O'Reilly Media for a conference title in 2004 (O'Reilly, 2005). O'Reilly considered the web as a platform, that empower users to create and manage content, and transform static websites into interactive ones, so that visitors could participate on-line and collaborate. Later on, O'Reilly offered a compact definition of Web 2.0 as "the business revolution in the computer industry caused by the move to the Internet as platform, and the attempt to understand the rules for success on that new platform" (O'Reilly, 2006). The characteristics of Web 2.0 are: rich user experience, user participation, dynamic content, metadata, web standards and scalability (Best, 2006). Characteristics, as openness, freedom and collective intelligence by way of user participation, can also be viewed as essential attributes of Web 2.0 (http://en.wikipedia.org/wiki/Web_2.0).

Web 2.0 has, at its heart, a set of at least six powerful ideas that are changing the way some people interact (TechWatch, 2007). The ideas are:

- Individual production and user-generated content
- Harnessing the power of the crowd
- Data on an epic scale
- Architecture of participation
- Network effects
- Openness

O'Reilly (2005) visualized Web 2.0 as a set of principles and practices that tie together (Figure 1)

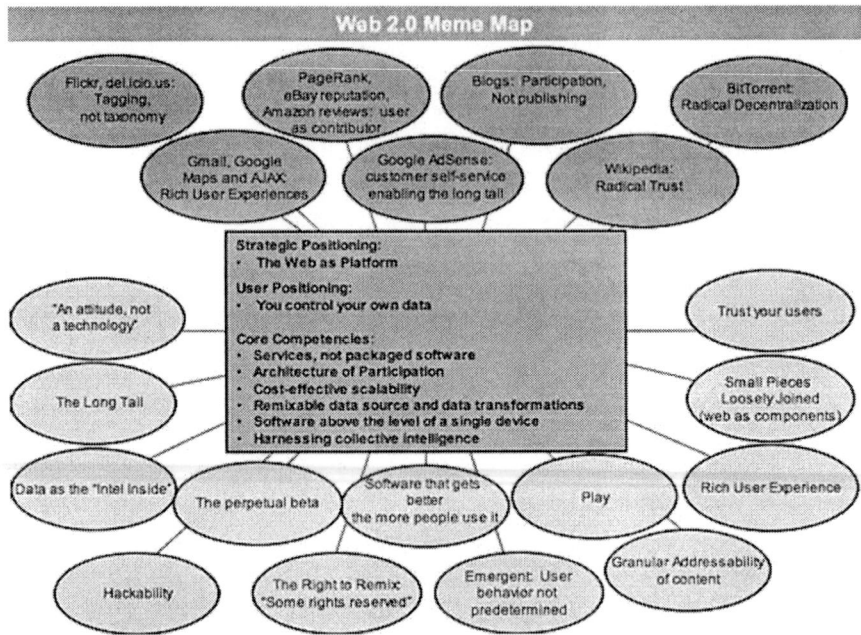

Figure 1. Web 2.0 Meme Map. Source: O'Reilly (2005), © 2005 O'Reilly Media, Inc. All rights reserved. Used with Permission.

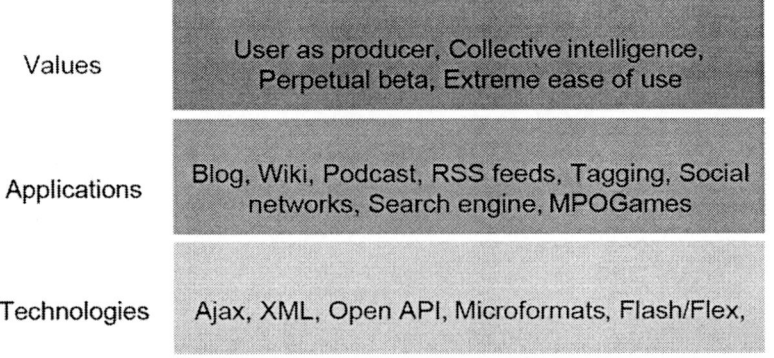

Figure 2. Operational description of web 2.0. Source: Osimo (2008). European Commission. Joint Research Centre. IPTS. *Technical Report EUR 23358 EN (2008)*.

Web 2.0 technologies and applications (e.g. tags, RSS, blogs, wikis, podcasts) has changed the way people search, find, read, gather, share, develop and consume information, as well as on the way people communicate with each other and collaboratively create new knowledge (Sigala, 2008). Web 2.0 also, can aid organizations and enterprises to enhance their businesses and sustain their competitive advantage (Gilchrist, 2007).

WIKIS

Wiki means "fast" in Hawaiian and the term is used to refer to websites that allow users easily add, delete and edit website content. A wiki is a collaborative website whose content can be edited by anyone who has access to it (Sigala, 2008). Ward Cunningham was the developer of the first wiki software, WikiWikiWeb (http://c2.com/cgi/wiki? WikiWikiWeb). WikiWikiWeb is described as the "home to an Informal History Of Programming Ideas as well as a large volume of material recording related discourses and collaboration between its readers.." The ease of access, intuitive interface and single repository – all of which contribute to the notion of a "living document" - make wikis an efficient and effective tool of mass collaboration (*www.spikesource.com/suitetwo/ downloads/web20white paper. pdf*). The only thing that a user needs to edit and read a wiki is a web browser, and this characteristic gives the wiki its great potential as a tool for online collaboration (Augar et al., 2004). The most popular wiki is the wikipedia.org website. Wikipedia.org is an online encyclopedia that is created and continually updated by its users.

A political wiki (http://openpolitics.ca/political%20wiki) is a wiki which can be used to:

- inform debate on public policy issues
- support deliberation via debate by edit
- enable participatory democracy

The core function of wikis, that is the ability for many people to change shared content easily, seems to match well with the goals of Direct Democracy and other initiatives that seek to make politics transparent (Makice, 2006).

In 2004 the Green Party of Canada (GPC) used, for the first time, a wiki to construct political positions. The Living Platform was created in order to motivate Canadians to participate in the annual update of their political platform. Political wikis include the Platform For Pittsburgh (http://Platform.For-Pgh.org) and the short-lived NOLA-Intel wiki, which published information about New Orleans in the early aftermath of Hurricane Katrina (Makice, 2006). Campaigns Wikia is another political wiki. (http://campaigns .wikia.com/wiki/Campaigns_Wikia). On the main page we can read the mission of the wiki: "It's time for politics to become more intelligent, and for democracy to really involve the people. Broadcast media tells you what to think and doesn't let you get involved. It's time to focus on what you need, what you care about, and the messages you want to get out.." GrassRoots (http://grassroots.wikia.com/wiki/Main_Page) is also a Wiki where political entities can interconnect on the issues, actions, projects, and events that occur throughout the range of geopolitical levels that people are a part of and make.

RSS (REALLY SIMPLE SYNDICATION)

RSS, is an abbreviation for "Really Simple Syndication." Software programs known as "Feed Readers" (a web-based environment) or "Aggregators" (similar to an email inbox) routinely check a user's "subscribed feeds" to see if any of those feeds have new digital content such as news, blogs or podcasts. If there is new or updated content, the digital content is retrieved and that content is presented to the user (www.spikesource.com/suitetwo/ downloads/web20whitepaper.pdf.)

The user subscribes to a feed by entering into the reader the feed's URL or by clicking an RSS icon in a web browser that initiates the subscription process. The RSS reader checks the user's subscribed feeds regularly for new work, downloads any updates that it finds, and provides a user interface to monitor and read the feeds. RSS readers enable Internet users to gather and read all new information that is customized to the user's profile within one consolidated message (Sigala, 2009).

The RSS formats were preceded by several attempts at web syndication that did not achieve widespread popularity. The basic idea of restructuring information about websites goes back to as early as 1995, when Ramanathan V. Guha and others in Apple Computer's Advanced Technology Group developed the Meta Content Framework. RDF Site Summary, the first version of RSS, was created by Guha at Netscape in March 1999 for use on the My.Netscape.Com portal. This version became known as RSS 0.9. The RSS-DEV Working Group, a project whose members included Guha and representatives of O'Reilly Media and Moreover Technologies, produced RSS 1.0 in December 2000. In July 2003, Winer and UserLand Software assigned the copyright of the RSS 2.0 specification to Harvard's Berkman Center for Internet & Society. In December 2005, the Microsoft Internet Explorer team and Outlook team announced on their blogs that they were adopting the feed icon first used in the Mozilla Firefox browser (http://en.wikipedia.org/wiki/RSS).

RSS allows new communication and interaction modes with information (Farmer, 2004). In RSS, both the communicator and the reader of information have control of the communication process, i.e. the former sends information only to those that users have selected to aggregate the RSS feed, while the latter selects from where and how to receive communication (Sigala, 2009).

RSS feeds help politicians to communicate with the general public about their positions on various issues and to communicate political statements on newsworthy issues (http://www.feedforall.com/politics.htm). Feeds can also help keep the community aware about political schedules, speeches and related events. Finally, RSS feeds may be created for blogs to communicate with a wider audience.

SOCIAL NETWORKING

Social network sites (SNS) like Facebook, MySpace, Friendster, Linkedin, Orkut, Cyworld, Bebo and Tribe are, perhaps, the best examples of O'Reilly's (2005) Web 2.0 environment (Valenzuela et al., 2008). There are hundreds of SNSs, with various technological affordances, supporting a wide range of interests and practices (Boyd & Ellison, 2007). Social network sites enable users to create their profile and invite others with similar profiles to take part in their online community (Sigala, 2009). Social network sites can be defined as "web-based services that allow individuals to (1) construct a public or semi-public profile within a bounded system, (2) articulate a list of other users with whom they share a connection, and (3) view and traverse their list of connections and those made by others within the system. The nature and nomenclature of these connections may vary from site to site" (Boyd & Ellison, 2007).

Social network sites may fulfill many of the promises of civic journalism. They deliver shared, relevant information to its users and provide a place for exchanging ideas (Merritt, 1998), media uses SNSs to help citizens connect to society and facilitate civic action (Rutigliano, 2007) and journalists and traditional news organizations can engage individuals, especially young adults, in public life (Valenzuela et al., 2008).

All of the major social network sites allow their members to form groups centred on almost any topic or theme and then identify and connect with others who have the same interest. To date, several thousand groups have been organized along a political theme. Social network sites, used as participatory tools, emerged in 2006 and established forms of two-way communication between parties or candidates and voters. Parties are attempting to exploit the free interactive communication means offered by social network sites; this is usually the 'Bulletin Board' or 'Wall.' This allows any visitor to post messages and the host to respond. Such features clearly offer the potential for participation in conversations either on the Wall of the party or leader, or across 'Walls.' However, parties use SNSs more to post extensive amounts of information about themselves, their policies and update links to favourable news items than to allow users to post conversations (Lilleker & Jackson, 2008).

The most prominently used SNS by U.S. congressional and gubernatorial candidates in 2006 was Facebook. Candidates or their campaign staff personalized the profile with everything from photographs to qualifications for office. Facebook members view these entries and register their support for specific candidates. They also received notifications every time one of their Facebook friends registered support for a candidate. Facebook displayed the number of supporters for each candidate and calculated the percentage of votes that candidate had in his or her race. 1.5 million members were connected either to a candidate or to an issue group (Williams. & Gulati, 2007).

TAGGING

A tag is a keyword that is added to a digital object (e.g. a website, picture or video clip) to describe it, but not as part of a formal classification system (Anderson, 2007). Tags are chosen by the user, not selected from a controlled vocabulary. So, tagging is also known as consumer-generated taxonomy. Tagging is used not only for saving and sorting a user's content but also for sharing content with others (Sigala, 2009). Social bookmarking systems allow users to create lists of 'bookmarks' or 'favourites,' to store these centrally on a remote service and in that way to share them with other users of the system (Anderson, 2007).

Del.icio.us and Yahoo's MyWeb allows collaborative tagging of shared website bookmarks, CiteULike and Connotea for references to academic publications, Flickr for photographs, YouTube for video and Technorati for weblog posts (Golder & Huberman, 2005). These sites include tag clouds that mean groups of tags from a number of different users of a tagging service. Tags are arranged alphabetically, with the most used or popular keywords shown in a larger font. This frequency information is often displayed graphically as a 'cloud' in which tags with higher frequency of use are displayed in larger text (Anderson, 2007; Sigala, 2009).

On 24 March 2009, CNN created what they claimed was the "largest word cloud in the free world" for that night's Anderson Cooper 360°. It was a word cloud of President Obama's address to the press earlier that day. (http://en.wikipedia.org/wiki/Tag_cloud).

In recent years tag clouds gained even more popularity because of their role in the search engine optimization of webpages. Properly implemented tag clouds make the website appear to search engine spiders more interlinked which tends to improve its search engine rank (http://www.software mastercenter.com/free-tag-cloud-generator-script.html.) Google (www.google. com) dominates the search engine market because they do not only rely only on pure syntax comparison but additionally exploit metadata about the structure of the World Wide Web. Google counts the number of links which point to a web page assuming that frequently linked pages are more important than seldom linked ones (Bielenberg, K. and Zacher, M. (2006). "What happens to your site when a dozen of your biggest supporters start tagging your blog and campaign site?" wonders Godin (2006).

PODCASTING AND ONLINE VIDEO

Podcasts are audio recordings, usually in MP3 format, which can be played either on a computer or on a wide range of handheld MP3 devices. Vidcast or vodcast are video podcasts that use the online delivery of video-on-demand clips and can be played on a PC, or again on a suitable handheld player (Anderson, 2007). Gil de Zúñiga et al. (2010) defined podcast as: "a digital audio or video file that is episodic; downloadable; programme-driven, mainly with a host and/or theme; and convenient, usually via an automated feed with computer software.." Podcasting refers to the uploading of audio and video files by users on websites. The most popular website for sharing such content with others is youtube.com (Sigala, 2009). The word 'podcast' was invented in 2004 and it combines the words iPod, a well-known music player and broadcast (http://www.oup.com/elt/catalogue/ teachersites/oald7/wotm/ wotm_archive/ podcast? cc=global).

A list of all the audio or video files currently associated with a given series is maintained centrally on the distributor's server as a web feed. Special software, "podcatchers," can automatically find and download the latest podcasts to a computer as soon as they becomes available. These files can also be downloaded to portable media players that can be taken anywhere, providing the potential for "anytime, anywhere" learning experiences (Sigala, 2009).

In July 2005, during the London bombings people on the Underground captured the events through picture and video phones. They posted these clips on websites and sent them to mainstream media sources, such as the BBC, which received over 1,000 still photos and 300 videos in the days following the explosions.

Podcasts make the relationship between political talk and online participation more and more interesting (Gil de Zúñiga et al., 2010). Cable companies are pitching politics on demand after trial runs in Colorado's 2004 U.S. Senate race and the 2005 governor's race in New Jersey, which allowed voters to order free clips of the candidates discussing issues. "In the latest creative wrinkle, politicians are podcasting -- White House hopefuls Gen. Wesley K. Clark, John Edwards and Sen. Bill Frist are among those regularly offering their downloadable ruminations -- and turning up on Flickr, MySpace, YouTube and other photo- and video-sharing Internet sites"(Barabac, 2006).

BLOGS

Blogs will be presented and discussed extensively in the next chapter.

Chapter 2

BLOGS

Barger (1997) used the term weblog for the first time and defined it as "a web page where a blogger 'logs' all the other web pages he finds interesting." The short form, "blog," was coined by Peter Merholz in 1999 who broke the word weblog into the phrase 'we blog' in the sidebar of his blog Peterme.com (Boyd, 2006). This word was quickly adopted both as a noun and as verb. *Opendiary, Live Journal* and *Diaryland* were the first programs allowing people unfamiliar with web-design to create blogs (Grieve et al., 2010).

Blog hosting services were introduced on August 23, 1999, when Blogger was launched by Pyra Labs, which was bought by Google 2003 (http://www.blogger.com). Blogger was credited for helping people popularize the format of blogs. As one of the earliest dedicated blog-publishing tools, with high popularity, it helped the spread of the term across the web and solidified the look of blogs (Boyd, 2006). In the main page of Blogger (http://www.blogger.com/tour_start.g) the term blog is defined as "A blog is a personal diary. A daily pulpit. A collaborative space. A political soapbox. A breaking-news outlet. A collection of links. Your own private thoughts. Memos to the world.." Drezner & Farrell (2004, p. 5) defined weblogs as "a web page with minimal to no external editing, providing on-line commentary, periodically updated and presented in reverse chronological order, with hyperlinks to other online sources," and Kolari et al. (2006, p.92) as "websites consisting of dated entries typically listed in reverse chronological order on a single page." Gumbrecht (2004) characterized blogs as a personal protected space where the author can communicate with others while retaining control.

The social definition of blogs asserts that they are low-threshold tools for Internet users to personalise and actualise content and information online and to present their views to a broad audience. Everyone can set up a blog in less than ten minutes with a minimal cost (Drezner & Farrell, 2004). Pedley (2005, p.95) mentioned "A major attraction of weblogs is their relative ease of construction/updating and the lack of the need or any special skills in web design or of HTML coding" and Jackson (2006, p. 294) wrote that with blogs "even technophobes could get online." Rosenbloom (2004) claimed "a blogger only needs a computer, internet access and an opinion." However, it takes effort to maintain an "active" blog. Activity depends on two parameters: the blogger, who needs to update the content regularly, and the readers who need to visit and interact often with the blog. It is a common phenomenon to abandon blogs soon after their creation (Hsu & Lin, 2008). Blogs help self-expression and self-empowerment, and are becoming a major source of information and communication for Internet users and have the power to engage people in collaborative activity, knowledge sharing, and reflection and debate (Blood, 2002; Efimova et al., 2005; Hiler, 2002; Punie & Cabrera, 2005). Blogs are not only employed in personal environments but also in organizations and enterprises (Kolbitsch & Mauer, 2006).

Weil (2003) gave the top 20 definitions for blogging and wrote. Blogging is...

- A form of unedited, authentic self-expression
- An instant publishing tool
- An online journal with freshly updated content
- Amateur journalism
- Something that will revolutionize the Web (think RSS feeds)
- A way to create community with your voters, er... readers (think 2,200 comments posted to the Dean for America blog in one day)
- An alternative to mainstream media (think InstaPundit by Glenn Reynolds and TalkingPointsMemo by Joshua Micah Marshall)
- A tool to teach students how to write
- A new way to communicate with customers (think Ray Ozzie, CEO of Groove Networks)
- A new form of knowledge management inside big companies
- A way for a bunch of navel-gazers to communicate with one another
- Something to keep you occupied when you're unemployed (more people than care to admit fit into this category... have you noticed?)
- A way to think and write in short paragraphs instead of a long essay (which no one has time to read anyway)

- Your email to everyone, as A-list blogger Doc Searls puts it (i.e., a way to stay in touch with family and friends)
- A silly word that's fun to say ("Gotta go blog now....")
- A way of writing with a distinct voice and personality (think Halley Suitt)
- Something to talk about at cocktail parties ("I blogged Seth Godin and he blogged me back...")
- A URL to add to your resume (as in TokyoTim, my 23-year-old son, who's working as an English teacher in Japan for a year)
- Something else to do with your mobile phone... think audio blogging and moblogging
- Something you don't want your mother to read: what my mother says about blogging

TYPES AND USE OF BLOGS

Regarding types of blogs Blood (2002) and also Herring et al. (2004) distinguished three types: personal journals, filters and notebooks. In personal journals, the content is internal by means that the blogger express thoughts and daily activities. The content of filters is external to the blogger: they 'filter' information from other sources on the web. Filters are pointing to content that the blogger finds interesting and the blogger normally provides commentary on it. Finally, notebooks may contain either external or internal content but are distinguished by longer, focused essays. The majority of blogs are "personal-journals in which authors report on their lives and inner thoughts and feelings" (Herring et al, 2004 p. 6). Entries in these weblogs document everything from "what the blogger had for lunch that day; movie and music reviews; descriptions of shopping trips; through to latest illustrations completed by the blogger for offline texts; and the like" (Lankshear & Knobel, 2003 p.2).

Another approach by Krishnamurthy (2002) classified blogs into four basic types according to two dimensions: personal vs. topical, and individual vs. community.

According to this approach, the personal journals like those found on LiveJournal.com are examples of quadrant I. A group of friends collaboratively blogging about personal matters is an example of blogs of quadrant II. 'Filter' blogs, which select and provide commentary on information from the web, are examples of quadrant III. Finally, MetaFilter - community weblogs fall into quadrant IV (Herring et al., 2004).

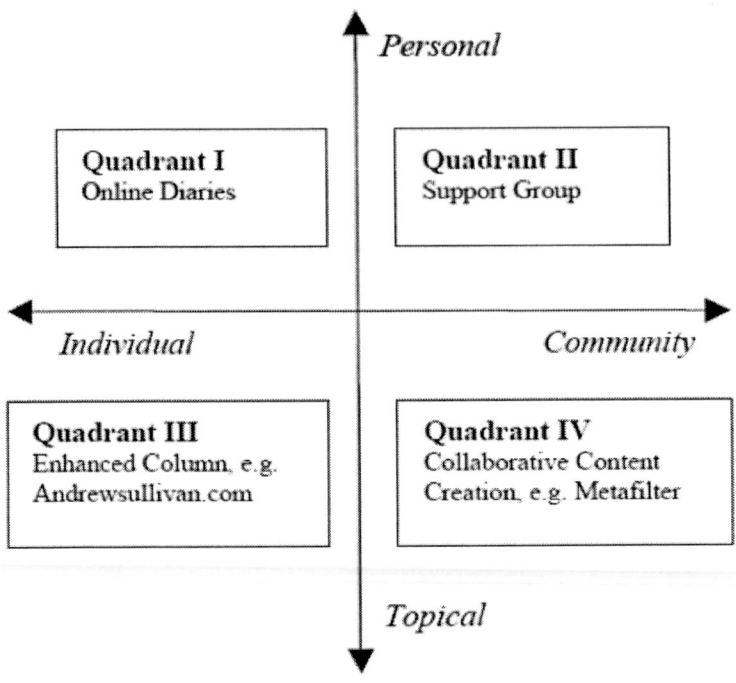

Figure 3. Types of blogs. Source: Krishnamurthy (2002).

As blogs are growing in popularity, businesses and organizations are looking for ways to exploit them. An organizational or corporate blog is a means of communication between organization/corporate and its public. Kelleher and Miller, (2006) commenting on corporate bloggers wrote, "These are people who blog in an official or semi-official capacity at a company, or are so affiliated with the company where they work that even though they are not officially spokespeople for the company, they are clearly affiliated." In the business world, blogs are considered as environments for knowledge sharing, (Festa, 2003) a unique opportunity for informal knowledge sharing as described by Kosonen et al. (2007) a "magic" formula for corporate communication (Jüch & Stobbe, 2005) and a potential for future profit (Lu & Hsiao, 2007). Companies like Microsoft and IBM encourage employees to embrace blogs (Du & Wagner, 2006) and Disney shares knowledge through an internal team blog (Guo et al, 2008).

Companies also use blogs in order to become recognized in the industry and to take direct feedback from customers, by allowing the public to make comments on blog posts (Hepburn, 2007). Robert Scoble , "The Scobleizer," as he's known to his daily readers, writes a Web log, or blog, posting comments on many topics however mostly, he writes about Microsoft. He is, in fact, a Microsoft employee. He gets feedback from tech-savvy readers on how to improve Microsoft products, and at times, he's even mildly critical of his employer. (http://www.fastcompany.com/magazine/81/blog.html)

Southwest Airlines, Marriot and Starwoods Hotels and Resorts maintain such official blogs (Dwivedi et al, 2007; Sigala, 2009). Sigala (2009) mentioned that Starwood has created its blog in order to communicate with its Preferred Guests and enhance their loyalty through the website www.thelobby.com. Moreover, companies can use selected blogs for advertisement. Companies such as Nike and Paramount utilise blogs as a new way to reach potential customers (Hsu & Lin, 2008). Amazon.com uses a blog in order to announce product enhancements and provide support (Guo et al., 2008).

In determining a blog for business use, Yap et al., (2005) proposed the RUBRIC model which reflects the six tests that make up the model for the blog's viability. RUBRIC stands for Reliability, Usability, Behavior, Reflection, Information, and Creativity. The model can be utilized in the form of a checklist, discussion points with developers and project supporters when the blog being developed.

At the core of the RUBRIC model is "reliability." Reliable sources include C-level executives, industry experts and strategists, and well-respected thinkers in the field. The second layer of this model is "ease of use." A graphic, table, text, and other navigation tools should be appropriately placed to reflect the mission, vision, and goal of the blog. Any decision on other design elements as pop-up windows, links to other sites etc, should thematically align with the business directives of the blog. The next layer is "behaviour" which is the appropriateness of the blog. Audience's perception should align with the corporate culture intended by the blog developers. When this behaviour is inconsistent across all blog pages and/or posting entries, then the "idea" of the blog crumbles. "Reflection" is the forth layer of the model and wraps the other three layers of the model as it is in this stage where there may be constant flux due to the many external factors affecting the blog's content, appearance, and navigation. It is important for a blog developer to be flexible in their product. Next, comes the "information" level. The blog information should always be current and relevant and reliable. The main concept of a blog is its ability to be practically instantaneous in its postings. When this inherent blog characteristic fails, the website as a whole becomes ineffective. The overall layer that surrounds the model is "creativity" and through in the form of the "spirit" of the developers and its writers. The uniqueness of a blog provides a lasting imprint to its readers. When readers use a blog's links to leave the blog without returning, then the site's "hook" is lost. When the blog is noted for its overall refined state, then its creativity edge is engaged.

Fuchs (2007) highlighted blogs' potential to advance cooperation and participation in social systems and Hoogenboom et al. (2007) their support to social learning processes.

In education, blogs can be used as an important tool for networking among teachers, students and parents. Blogs can also be used as a classroom management tool. Teachers can post assignments to students and students can share their learning experiences and express their thoughts to the instructor (Maag, 2005). Blogs can facilitate extended discussions beyond class meetings (Betts & Glogoff, 2004), increase students' intellectual exchange (Williams & Jacobs, 2004) and finally enhance their studies (Kim et al., 2008). According to Duffy and Bruns, (2006) potential benefits of blogs' use in education are (i) Promotion of critical and analytical thinking. (ii) Promotion of creative, intuitive and associational thinking (iii) Promotion of analogical thinking (iv) Potential for increased access and exposure to quality information. (v) Combination of solitary and social interaction. Librarian blogs serve as new information channels both for the library and information science community and for the general public (Bar-Ilan, 2007).

BLOG CHARACTERISTICS

Even though blogs can take a number of forms, they tend to have a number of characteristics in common. Blogs are set up and controlled by the author, display blog postings in reverse-chronological order or are arranged by subject. Each entry has a timestamp so that the reader knows when it was posted and to alert readers to the "currency" or "timeliness" of the log. They are updated regularly; some bloggers update several times a day, while others may update every few days or once a week with new material. Blogs usually tend to cover a very specific subject area, provide a personal viewpoint and facilitate critical feedback by giving their readers the opportunity to comment on blog postings, or to contact the author directly (Lankshear & Knobel, 2003; Pedley, 2005).

A blog is usually made up of the following components:

- Post Date — date and time the post was published
- Category — category the post is labelled with (can be one or more)
- Title — main title of the post
- Body — main content of the post
- Trackback — links back from other sites
- Comments — comments added by readers
- Permalink — the URL of the full, individual article
- Footer — usually at the bottom of the post, often showing post date/time, author, category, and stats such as number of reads, comments or trackbacks (Duffy and Bruns, 2006)
- Syndication feeds (Danah, 2006)
- Blogroll — a list of links of other blogs that the blog author reads or affiliates with.

Blogging tools enable between-blog interactivity, building up in that way the "blogosphere" whereby social networks among bloggers are created (Sigala, 2009). Du and Wagner (2006), www.scripting.com and Sigala (2009) identified the following characteristics of blogs:

- **Personalized:** blogs are designed for individual use and their style is personal and informal. Multi-person weblog is also possible through collaboration; for example Blogger.com offers a "team blog" collaborative feature enabling multi-person weblog.
- **Web-based:** blogs can be updated frequently and are easy to access by simply using a web browser.
- **Community-supported:** Weblogs can link to other weblogs and websites (e.g. photos, videos, web-texts), enabling the synthesis and linkage of ideas from different users, and so, stimulating knowledge generation and sharing between bloggers.
- **Automated:** Blogging tools are easy to create and maintain without the need of writing HTML code or program, so bloggers can solely concentrate on the content.

BLOGGERS

Bloggers are not a homogenous group, but they are an educated and affluent one. According to Technorati.com (2008) three out of four U.S. bloggers are college graduates, and 42% have attended graduate school. They skew male, and more than half have a household income over $75,000. Bloggers are also experienced. Half of bloggers are on their second blog, and 59% have been blogging for more than two years.

In Greece, forty percent of the population uses the Internet. Percentages are higher among young people and men. Searching for information, sending emails, downloading, playing games, chatting, and online buying, these are their reasons for using the internet. Using a sample of 1367 bloggers (Karampasis, 2007), found out that blogging started to expand during 2002-2003 in Greece. There exist 9510 blogs written in Greek, but only 4639 of them are active. The average Greek blogger is around 30, with a college education. S(he) uses DSL connection. Blogs receive less than 100 visits daily, while they do not have any advertisements and they discuss multiple subjects. Bloggers are 64% male. 65% of the bloggers live in Athens (53.1%) and Thessaloniki (12.4%). Eleven percent are residents living abroad. Motives for blogging are: keeping a diary, experimenting, taking action while being anonymous, or for the creation of a community. Personal interests, art and culture, and entertainment are the main subjects throughout Greek blogs. News and politics blogs are rarer. Thirty eight percent of the bloggers consider blogging to be a form of journalism, while 51% does not.

BLOGS AND JOURNALISM

Usually blogs are maintained by individuals to put forward their personal thoughts, however there are many blogs written by experts within a particular subject. Blogs are often perceived as powerful because they allow millions of people to easily publish and share their ideas, and millions more to read and respond (www.technorati.com) and because they are a good way of circulating new ideas (Baker & Green, 2005).The fact is that blogs have reshaped the web, impacted politics, and shaken up journalism, and enabled millions of people to have a voice and connect with others (http://www.blogger.com/tour_start.g). A question that arises is whether blogging is journalism.

Blogging and journalism aren't the same, but they achieve the same results. A blogger writes out of passion, out of an extreme interest for a particular topic (http://chris.pirillo.com/are-bloggers-journalists-are-blogs-new-journalism/). Definitely, not everyone who keeps an on-line journal is a journalist. Andrews (2003) wrote: "it is fair to say the vast majority of blogging does not qualify as journalism. If journalism is the imparting of verifiable facts to a general audience through a mass medium, then most blogs fall well short of meeting the standard." However, blogs are a form of journalism and are changing the way in which journalism is practiced today. Lasica (2002) and Pedley (2005) characterized blogs as "amateur" journalism." The most vital and moral dispatches on the Web are being created by amateurs," mentioned Lasica (2002) and Pedley (2005) wrote "This can be seen as part of their strength but also as one of their main weaknesses." Wall (2004) characterized them as "informal journalism" and "black market journalism" because they are not operating within the standard set of rules of formal journalism, and she highlighted the fact that blogs are attempting to bypass traditional media news channels. As blogs have the capability to report news without the constraints of censure or the pressures of advertising they can offer deeper analysis, based upon diverse range of sources and contributing citizen commentators, which is not possible through mainstream media (Kenix, 2009). However diverse opinions exist. Allan (2002) mentioned that news blogs, unlike mainstream media, are generally not objective or detached but opinionated and personal and this is their main weakness. The author "may well have a particular bias or point of view that they want to get across, and so it is important to take that into account when reading their content" (Pedley, 2005 p.96). On the other side (Kenix, 2009 p.790) wrote "the recognition of ideological forces in the news has led many to believe that mainstream journalism may not able to serve its democratic function in society."

There is an open argument between bloggers and journalists about how much faith to place on blogs (Johnson et al., 2007). Even though blog users are aware that bloggers are opinionated they trust blogs and this gives blogs the power to bring about social change (Banning & Trammell, 2006; Johnson & Kaye, 2004; Johnson et al., 2007). Especially politically interested Internet users rely more on blogs than on any other news source for news and information and judge blogs as more credible than either portal or mainstream news sites (Johnson & Kaye, 2007). Controversial opinions also exist. According to Koh et al., (2005, p. 2) " bloggers are often criticized for not being "professional journalists," with the assumption being that they have neither the training nor the judgment necessary to present news and opinions that could affect public opinion."

Another question that arises in any discussion about blogs and journalism, is: Which blogs? Blood (2002) mentions that four types of blogs are most frequently cited: (i) Blogs written by journalists; (ii) Blogs written by professionals about their industry (iii) Blogs written by individuals at the scene of a major event; (iv) Blogs that link primarily to news about current events.

Journalists regard blogs as alternative sources of news and public opinion (Lasica, 2001) even thought they view blogging sceptical (Williams & Jacobs, 2004) and perhaps a threat to their profession (Thompson, 2003). Many journalists keep their own blogs and very few of them are paid by their organizations to do so. The majority of journalists do not keep blogs (Andrews, 2003). However journalists read blogs and they read blogs much more than the general public (MacKinnon, 2007). According to Trammell & Keshelashvili (2005), in 2005, 51% of journalists read blogs and 53% get story ideas or sources from them. Farrell & Drezner (2008) claimed that over 83% of journalists also read blogs, and 43% use them on a weekly basis. Another survey by Brodeur and MarketWire in 2008, shows that over 75% of journalists use blogs to get ideas for stories. 70% of reporters check a blog list on a regular basis, 21% of reporters spend over an hour per day reading blogs and 57% of reporters read blogs at least two to three times a week. Almost 30% of journalists in the survey say they have their own blog (Thaeler, 2008). Regarding the reasons that journalists find blogs useful the top choices at their survey are: 1. "A way to find out about emerging stories sooner than I would otherwise" 2. "As a general source of story ideas" 3. "As a source of information that I can't find elsewhere" and "As a gauge of China's popular culture "pulse" 4. "To locate interesting people to interview" 5. "As a gauge of the sentiments of China's people on various issues" 6. "Because I want to know what certain bloggers have to say on a topic."

Robinson (2006, p.79) claimed that journalists' blogs are the corporate answer to independent bloggers and she mentioned "They are a way for journalists to reclaim journalism – and its standards – online, even through a postmodern entity originally created to defy those traditional norms." Newspaper blogs also exist, which are very popular. Professionally written blogs now appear at *MSNBC, Slate, the Washington Post*, the *Christian Science M*onitor, the *Seattle Times* (Gill, 2004). Pohlig (2003: 25) wrote on this ."... are hugely popular, largely because they allow the reader to see the journalist as a human being, connecting with them without the stiff, imperial we voice that turns so many younger people off."

Blogs written by professionals about their industry have emerged as a new connection between professionals and the public and they convey a complete version of the news about their profession. Blood (2003) claimed that "Their commentary, done with integrity, can be a great source of accurate information and nuanced, informed analysis, but it will never replace the journalist's mandate to assemble a fair, accurate and complete story that can be understood by a general audience."

ETHICS IN BLOGGING

One of the most significant charges against bloggers is that they do not keep to ethical standards of news reporting (Koh et al., 2005). On Cyberjournalist.net one can read "Some bloggers recently have been debating what, if any, ethics the Weblog community should follow. Since not all bloggers are journalists and the Weblog form is more casual, they argue they shouldn't be expected to follow the same ethics codes journalists are. But responsible bloggers should recognize that they are publishing words publicly, and therefore have certain ethical obligations to their readers, the people they write about, and society in general." So, Cyberjournalist.net has produced a blogger's code of ethics by modifying the Society of Professional Journalists Code of Ethics for the Weblog world. The code gives responsible bloggers a set of ethical obligations to work with. The basic principals are: Honest and Fair, accountability and minimizing harm done to others by blogging. "Bloggers who adopt this code of principles and these standards of practice not only practice ethical publishing, but convey to their readers that they can be trusted" (www.cyberjournalist.net/news/000215.php).

The Bloggers' Code of Ethics Is: Be Honest and Fair

Bloggers should be honest and fair in gathering, reporting and interpreting information.

Bloggers should:

- Never plagiarize.
- Identify and link to sources whenever feasible. The public is entitled to as much information as possible on sources' reliability.
- Make certain that Weblog entries, quotations, headlines, photos and all other content do not misrepresent. They should not oversimplify or highlight incidents out of context.
- Never distort the content of photos without disclosing what has been changed. Image enhancement is only acceptable for for technical clarity. Label montages and photo illustrations.
- Never publish information they know is inaccurate -- and if publishing questionable information, make it clear it's in doubt.

- Distinguish between advocacy, commentary and factual information. Even advocacy writing and commentary should not misrepresent fact or context.
- Distinguish factual information and commentary from advertising and shun hybrids that blur the lines between the two.

Minimize Harm

Ethical bloggers treat sources and subjects as human beings deserving of respect. Bloggers should:

- Show compassion for those who may be affected adversely by Weblog content. Use special sensitivity when dealing with children and inexperienced sources or subjects.
- Be sensitive when seeking or using interviews or photographs of those affected by tragedy or grief.
- Recognize that gathering and reporting information may cause harm or discomfort. Pursuit of information is not a license for arrogance.
- Recognize that private people have a greater right to control information about themselves than do public officials and others who seek power, influence or attention. Only an overriding public need can justify intrusion into anyone's privacy.
- Show good taste. Avoid pandering to lurid curiosity. Be cautious about identifying juvenile suspects, victims of sex crimes and criminal suspects before the formal filing of charges.

Be Accountable

Bloggers should:

- Admit mistakes and correct them promptly.
- Explain each Weblog's mission and invite dialogue with the public over its content and the bloggers' conduct.
- Disclose conflicts of interest, affiliations, activities and personal agendas.

- Deny favored treatment to advertisers and special interests and resist their pressure to influence content. When exceptions are made, disclose them fully to readers.
- Be wary of sources offering information for favors. When accepting such information, disclose the favors.
- Expose unethical practices of other bloggers.
- Abide by the same high standards to which they hold others.

Another approach was done by Koh et al. (2005). Based on the literature on Internet ethics (e.g., netiquette and nethics), blogging ethical codes, and journalism ethics, they identified four ethical principles relevant to blogging. They are: truth telling, accountability, minimizing harm, and attribution. Truth telling includes underlying concepts such as honesty, fairness, equality and completeness in reporting. Accountability involves being answerable to the public, honesty in one's work, revealing conflicts of interest, and bearing consequences of one's actions. The third principle is minimizing harm (done to others) by blogging. It includes issues of privacy, confidentiality, flaming, consideration of other peoples' feelings, and respecting diverse cultures and underprivileged groups. Attribution involves issues such as plagiarism, honouring intellectual property rights, and giving proper credit to sources.

POLITICAL BLOGGING

Political blogs are a new form of political expression and participation and have the potential to shape politics and political discourse (McKenna & Pole, 2004; Wallsten, 2005). The reference point for the political impact of blogs was the 2004 U.S. presidential campaign (Albrecht et al., 2007). Thereof blogs are used as a source of information for the purpose of political debate, to mobilize supporters and funding (Kerbel & Bloom, 2005) and to influence the news, political and policy agenda (Jackson, 2006). Blogs are also featured in papers regarding their political activism and influence (Su et al., 2005). Political blogs affected real world events (Drezner & Farrell, 2004) and have been described as an alternative channel for the distribution of information, as a mobilization tool due to their ability to spread news very quickly (Albrecht et al, 2007; Kahn & Kellner, 2004), as a credible source which provides depth and thoughtful analysis (Andrews, 2003; Johnson & Kaye, 2004; Regan, 2003), and as "gatekeepers of information and news" (Pedley, 2005 p. 97).

Moreover, blogs serve as a medium and tool for political change and many times are creating the media agenda (Cornfield et al., 2005). Gil de Zúñig et al. (2010 p.37) wrote " blog communities present avenues for individuals to be part of traditional political participation activities while also providing new online opportunities for the exchange of political perspectives and mobilization into action" and Herring et al. (2004) claimed that political blogs as possessing a "social-transformative, democratizing potential" and Walker (2007) called blogs "a soapbox" and highlighted their attractiveness for voicing political messages.

BLOGS AND MAJOR NEWS

Blogs – both political and personal commentary –were multiplied exponentially after 9/11, as people used them to express their political awareness (McKenna & Pole, 2004) and created a phenomenon known as war blogs (Gill, 2004). There is an expanding number of situations in which blogs have exercised an important influence over how politics is practiced and policy is developed and have broken major news sometimes contradictory to the point of view of the mainstream media (Jackson 2005; Lankshear & Knobel 2003; Sroca, 2006). "The interest within the political sphere on bloggers is that they are a potential alternative to the traditional media" mentioned Pedley (2005, p.295). Moreover, blogs have the advantage of speedy publication and there are many paradigms that political blogs provide that is not readily available elsewhere.

In December, 2002 bloggers created a protest against Trent Lott who supported Strom Thurmond's segregationist stance in the 1948 presidential election at the 100th birthday of Senator Strom Thurmond (Bloom 2003). When Trent Lott, on December 5, began his tribute, he said, "I want to say this about my state. When Strom Thurmond ran for president, we voted for him. We're proud of it. And if the rest of the country had followed our lead, we wouldn't have had all these problems over the all these years either." Even thought the mainstream press ignored the story, the continued attention of bloggers, Josh Marshall, who is an American Polk Award-winning journalist, and "Atrios" kept the story alive and drew media attention. On December 9, the conservative blogger Andrew Sullivan stated on his blog that "Trent Lott Must Go." On, December 10, Howard Kurtz of the *Washington Post* addressed the speech in his column. On, December 10 the *Post* and the *New York Times* wrote that there is evidence that Lott had used the exact same phrase in

November 1980 when introducing Thurmond at an event. On Friday December 20, Lott resigned as Majority Leader (Gill, 2004).

Bloggers, also echoed the suspicion that "President Bush was using a listening device for assistance during the presidential debate and worries over the validity of voting machines" (Su et al. 2005) and bloggers questioned the authenticity of documents held by CBS regarding President Bush's military service in the Texas National Guard in an incident known as "RatherGate" (Williams et al., 2005)

In November 2005, bloggers in Germany found that the picture used for the claim of the social marketing campaign had already been used by some Nazis in 1935 (http://www.spreeblick.com/2006/01/27/you-are-deutschland-too-just-kidding/). In the US-led invasion of Iraq, blogs posted photos before other media and sometimes contradicted the point of view of the mainstream media, like CNN (Lankshear & Knobel, 2003). The famous contemporary blogger Iraqi Salam "gave outsiders a dose of the larger unexpurgated reality as the bombs exploded overhead in Baghdad," and Hossein Derakhshan became the first blogger to be jailed for "undermining national security through cultural activities" (Kahn, R., & Kellner., 2004, p.90).

Sometimes bloggers use the uncensored nature of the internet to bypass state controlled news media (http://en.wikipedia.org/wiki/Political_blog). In liberal democracies citizens have the right to criticize the government, but in other countries it's a serious criminal charge and bloggers may find themselves persecuted. For example, in Iran, Mohamad Reza Nasab Abdolahi, had published an open letter to Ayatollah Ali Khamenei on his blog, Webnegar, and found himself facing six months in jail and a fine of one million rials. When his pregnant wife wrote about this on her own weblog, she too had her computer seized and was taken to prison (Connor, 2005). In Cairo-Egypt, Police had detained Rami Siyam a blogger who posted criticism of the government, on his blog Ayyoub. No reasons have been given for Mr Siyam's detention. Human rights groups have accused Egypt of eroding freedom of speech by arresting several bloggers (BBC News, 2006)

Another interesting phenomenon about blogging are the "bridge blogs." The term "bridge blog," was first used in 2004 in order to describe group of bloggers whose blogs act as "bridges" between their home country and a wider global audience. An early "bridge blogger" is exiled Iranian blogger Hossein Derakhshan, who in 2002 started blogging in English, summarizing what Iranian bloggers were writing in Persian. The emergence of such bridge bloggers around the world inspired the creation of the website Global Voices Online (GlobalVoicesOnline.org), whose international team of bloggers provides a daily selection of the best "bridge blogging" from around the non-Western world. (MacKinnon, 2007).

BLOGS AND POLITICAL PARTIES

Political parties use blogs in order to shape the political agenda, generate resources, mobilize support and reach out to their constituency (Trammell et al. 2006). According to Chadwick (2008) primary and presidential campaigns in the United States "saw the emergence of campaigning model based on online venues loosely meshed together through automated linking technologies, particularly blogs." Jackson (2006, p.295) suggest that "during an election campaign a weblog is a means for a party to promulgate its policies through a virtual network of political bloggers." During election campaigns parties and candidates blogs are used for three main functions: diffusing information to internal audiences, building up a volunteer base, and the agenda setting of the mainstream media (Bloom & Kerbel, 2005). However, a few of the political blogs have been run by political candidates (McKenna & Pole, 2004). Albrecht (2007, p.504) claimed that blogs "offer forms of communication that allow political actors to bypass established media practices." An example of a highly successfully blog is Howard Dean's Blog for America. The blog was used to mobilize supporters and funding, even though Howard Dean's candidacy was not successful (Albrecht et al., 2007; Kerbel & Bloom, 2005; McKenna & Pole, 2004).

Blogs have also raised money for candidates in elections in some congressional districts in USA. In Ohio, for example, even though Paul Hackett lost in elections to the Republican Jean Schmidt, his contests had been far more competitive than expected. In South Dakota politics, Jon Lauck and Jason Van Beek were credited with aiding John Thune's victory over Senate Minority Leader Tom Daschle, D-S.D. Thune's campaign reportedly paid both bloggers, and Thune later hired Van Beek as a Senate aide (Glover, 2006). According to Albrecht et al. (2007) paradigms that blogs use as campaign instruments were the 2004 presidential election, the 2005 U.K. general election, the 2005 Danish parliamentary election, the 2005 New Zealand general election, the 2005 German Bundestag election, the 2007 French election and the 2007 Australian Federal Election. At recent presidential elections in the USA, Barack Obama maintained a blog at his webpage (http://www. barackobama.com).

Obama's team showed a real understanding of online media to campaign and raise funds for the candidate. Barack Obama and his team ensured that his campaign had a presence on the major social networks. This meant pages on Facebook and MySpace as well as networks like BlackPlanet.com and even LinkedIn. On each of these sites he has managed to gather a big following - 500,000+ on both MySpace and Facebook, many more than John McCain could manage. Obama used also Twitter, Flickr, and a blog during his campaign. As with his social network activities, the blog and Twitter accounts were regularly updated, providing new content for search engines to crawl, and ensuring that it was worthwhile for people to keep coming back to check for updates (http://econsultancy.com/). Supporters wrote more than 400,000 blog posts on the MyBO Web site (Edelman, 2009).

Investigating political blogs for campaigns, Garrett (2004) (in Trammell et al. (2006)) mentioned that bloggers write posts in a personal voice, update the blogs several times a day, encourage and moderate comments, offer hyperlinks to internal and external sources, and other blogs, and call the readers into action. Graf & Darr (2004) pointed out that blogs appear to play an increasingly important role as a forum of public debate, with knock-on consequences for the media and for politics. "In just a few years they have become a finger in the eye of the mainstream media and a closely watched forum of political debate. Political blogs have exposed lapses in mainstream media coverage, chastened reporters with the fear of an angry online response to sensitive stories, and at times set the media agenda. Political blogs have also been influential in raising money for political candidates and pushing select races into the national spotlight" (Graf, 2006; p.3)

Graf (2006) tried to investigate the audience of political blogs and undertook an online survey. He used a sample of 7,863 people and they mentioned that "the minority of our sample has at least seen a political blog. About 40% of everyone we sampled said they had looked at a blog 'that discusses politics or current events'at least once in the past month. Another 7% said they visited "several times a week." However, only 9% of our sample respondents said they visited political blogs "almost every day." The findings of the survey saw that the daily audience for political blogs is fairly small however. Bloom (2003) also commented about visitors of political blogs in the USA, and mentioned that a high percentage was political reporters, politicians and policy makers: key opinion formers. Taking this in to consideration Jackson (2006, p.296) mentioned: "This can give political bloggers a disproportionate influence, based on the type of blog visitor, and not just the number of blog visitors. Therefore, elite bloggers can act as a "focal point" encouraging influential visitors to congregate around them. To influence the news, political and policy agenda, political actors need to attract an "A" list audience to their weblog."

In Greece, where the ratio of internet users is relatively small, there is however an expanding portion of bloggers who comment regularly and have the power to a certain degree and in certain circumstances to trigger off political movements. The posting of opinions on the Internet can be considered an expression of activated public opinion in contradiction to public opinion, which is recorded through surveys and concerns the wider public.

BLOGS HYPERLINKING

By definition, blogs link to other sources of information; usually to other blogs. The most important difference between blogs and more traditional media "is that blogs are networked phenomena that rely on hyperlinks" (Drezner & Farrell, 2004; p.5) and on this depend their success (Williams & Jacobs, 2004). Blogs depend upon hyperlinks not only to boost attention to their own blog, but to also ensure that users can be quickly led to relevant information (Kenix, 2009). Blogging tools provide appropriate features for managing blog interactivity and for promoting the creation of social networks among bloggers (Du & Wagner, 2006). Sigala (2008) mentioned, "blogs create and maintain strong online communities through their social ties- tools such as blogrolls, permalinks, comments and trackbacks."

A "blogroll" is a list of blogs that many bloggers maintain. The blogroll occupies a permanent position on the blog's home page and is the list of blogs that the blogger frequently reads or especially admires and thus offers links to these blogs. Blogrolls evolved early in the development of blogs and serve as a navigation tool for blog readers to find other blogs with similar interests (Marlow, 2004). Commenting on this, Drezner & Farell (2004, p.7) wrote: "Blogrolls provide an excellent means of situating a blogger's interests and preferences within the blogosphere. Bloggers are likely to use their blogrolls to link other blogs that have shared interests." Blogrolls are also great traffic driving tools. "With each blogroll that your blog is listed on comes the possibility that readers of that blog will click on your link and visit your blog" (http://weblogs.about.com /od/partsofablog/qt/WhatIsaBlogroll.htm). Albrecht et al. (2007, p. 506) referred to this form of interactivity between blogs as the "connectedness of weblogs."

Interactivity can be also achieved with permalinks and comments. A permalink (permanent link), is a URL that points to a specific blog entry after it has passed from the front page to the archives. An entry in a blog with many entries is accessible from the site's front page for only a short time. Visitors who store the URL for a particular entry often find upon their return that the desired content has been replaced by something new. Prominently posting permalinks is a method employed by bloggers to encourage visitors to store a more long-lived URL (the permalink) for reference. Permalinks frequently consist of a string of characters which represent the date and time of posting, and an identifier which denotes the author who initially authored the item or its subject. Crucially, if an item is changed, renamed, or moved within the internal database, its permalink remains unaltered, as it functions as a magic cookie which references an internal database identifier. If an item is deleted altogether, its permalink can frequently not be reused (http://en.wikipedia.org).

Comment tools allow readers to express their opinions on a weblog entry and permalinks allow weblogs link posts with one another. Comments are "reader-contributed replies to a specific post within the blog" (Marlow, 2004, p.3). Comments' system is implemented as a chronologically ordered set of responses and is the key form of information exchange in the blogosphere (Drezner & Farrell, 2004; Mishne & Glance 2006). "Posting volume would be a key determinant of content value" claimed Lu & Hsiao (2007, p. 346).

At last are trackbacks and pingbacks. Trackback is a citation notification system (Brady, 2005). "Trackbacks" introduced by Movable Type in 2001, and made formally invisible connections visible. Trackbacks enables bloggers to determine when other bloggers have written another entry of their own that references their original post. "If both weblogs are enabled with trackback functionality, a reference from a post on weblog A to another post on weblog B will update the post on B to contain a back-reference to the post on A" (Marlow, 2004). A pingback is an automated trackback. "Pingbacks support auto-discovery where the software automatically finds out the links in a post, and automatically tries to pingback those URLs, while trackbacks must be done manually by entering the trackback URL that the trackback should be sent to" (http://codex.wordpress.org/ Introduction_to_Blogging#Pingbacks).

A-LIST BLOGS

There are millions of individual blogs, but within any community, only a few blogs attract a large readership (Wagner & Bolloju, 2005). "The vast majority of blogs are probably only read by family and friends, there are only a few elite blogs which are read by comparably large numbers" wrote Jackson (2006, p.295). Herring et al. (2004) also claimed that the most discussions of the blogosphere focus on an elite minority of blogs. These blogs are the most known and regularly linked by others. According to Trammell & Keshelashvili (2005, p. 968) their authors manage to create a persona, making themselves a "celebrity" among the community of bloggers. These blogs are referred as "A-list." "A-list blogs—those that are most widely read, cited in the mass media, and receive the most inbound links from other blogs—are predominantly filter-type blogs, often with a political focus. The A-list appears at the core of most characterizations of the blogosphere" wrote Herring et al. (2005).

Many bloggers desire a wide readership; and the most reliable way to gain traffic to their blog is through a link on another weblog (Blood, 2002). Drezner & Farrell (2004, p.7) mentioned "when one blog links to another, the readers of the former blog are more likely to read the latter after having clicked on a hyperlink than they would have been otherwise. If they like what they read, they may even become regular readers of the second blog." In that way "blogs with large numbers of incoming links offer both a means of filtering interesting blog posts from less interesting ones, and a focal point at which bloggers with interesting posts, and potential readers of these posts can coordinate" (Drezner & Farrell, 2004; p.13). Less prominent bloggers contact one of the large 'focal point' blogs, to publicize their post when they have an interesting piece of information or point of view that is relevant to a political controversy. On the one hand, this leads the readers to 'focal point' blogs, as they know that they will find links to many interesting stories, and on the other hand leads bloggers to send posts to focal point blogs as they know that they are likely to find more readers. Based on this and the lognormal distribution of weblogs, in a given a political issue, the media only needs to look at the top blogs to obtain a "summary statistic" about the distribution of opinions (Drezner & Farrell, 2004). In this vein Adamic & Glance (2005, p.2) noted: "Because of bloggers' ability to identify and frame breaking news, many mainstream media sources keep a close eye on the best known political blogs." At another research Park & Jankofski (2008) investigated hyper linking of citizen blogs in South Korean politics and stated that "If there is an increasing frequency of neighbour links directly flowing through the blog of a politician, it may indicate the politician's role as the online community leader as well as the information hub for the community" (p.64). Drezner & Farrell (2004) found out that the median blogger has almost no political influence as measured by traffic or hyperlinks and they highlighted "This is because the distribution of weblinks and traffic is heavily skewed, with a few bloggers commanding most of the attention. This distribution parallels the one observed for political websites in general. Because of this distribution, a few "elite" blogs can operate as both an information aggregator and as a "summary statistic" for the blogoshpere." The same was claimed by Jackson (2006, p.296) who wrote "elite bloggers can act as a "focal point" encouraging influential visitors to congregate around them. To influence the news, political and policy agenda, political actors need to attract an "A" list audience to their weblog."

SOCIAL NETWORKING ANALYSIS

Social network analysis (SNA) is "a discipline of social science that seeks to explain social phenomena through a structural interpretation of human interaction both as a theory and a methodology" (Marlow, 2004 p.2). Its starting point is the premise that social life is created primarily and most importantly by relations and the patterns formed by these relations (Marin & Wellman, 2010). The goal of SNA is to identify "who the key actors are and what positions and actions they are likely to take" (Krackhardt, 1996, p.161). According to Balancieri et al. (2007) SNA is rooted in the concepts of nodes and connections. Nodes are the social actors and can be persons, groups, organizations, nations, communities, offices, blogs and so on and "connections" refer to channels of communication, (Balancieri et al., 2007; Martino & Spoto, 2006). The attributes of social actors, along with the properties of relationships between social actors, such as the nature, intensity, and frequency of the relationships, have important implications to the social structure. SNA methods have been employed to study organizational behaviour, inter-organizational relations, citation patterns, computer mediated communication, and many other areas (Chau & Xu, 2008). Blogs facilitate members' social interactions and provide conversation (Nardi et al., 2004; Herring et al., 2005) and in this vein, SNA can be used as a research vehicle to investigate the structural patterns of blogging communities.

A Social Network can be represented in three ways: the first one is by giving a simple list of all the elements taken from the set of social actors, and the list of the pairs of elements that are linked by a social relationship of some kind. The second has a form of matrix. If two social actors I and J have a relation then 1 is placed at the cell (i,j), otherwise 0 is placed in this cell. Finally, a description of a Social Network may have a form of a graph where social actors can be represented by nodes, and the connections with each other can be represented by edges between these nodes. If the graph is directed, each interaction describes a one-way association between two social actors. In this case, the in-degree of a node is the number of incoming links and the out-degree is the number of outgoing links. Graph Theory's approach is crucial as it denotes the structural properties of a network and provides a tool to quantify and measure some properties of the network (Marlow, 2004; Martino & Spoto, 2006).

"Blogs are a form of social hypertext, functioning as a one-to-one mapping between a network of web pages and a network of people, which can be represented as a social network and from which communities can emerge" wrote Chin & Chignell (2007). Blog communities emerge from interlinking between them (Efimova & Hendrick, 2005; Efimova et al., 2005). SNA exploits interlinking between blogs and examines the roles and behaviour of nodes on other nodes in the network, and on the whole network (Chin & Chignell, 2007). Park & Jankowski (2008, p. 60) claimed "The configuration of link networks themselves can be a source conveying useful overall information about the (hidden) online relationship of communication networks in interpersonal, inter-organisational, and international settings."

Chapter 3

ANALYSIS OF BLOGS COMMUNICATIONS PATTERNS

METHODOLOGY

This research serves two purposes. On the one hand, it makes a hyperlink analysis of the incoming links of the recorded blogs using Social Networking Theory (SNT) and Social Networking Analysis (SNA), and on the other, it studies the communication patterns among blogs. It attempts to portray the communication patterns generated among Greek political blogs through the linkages from blogrolls. This chapter records and analyzes blogroll hyperlinks interconnections. It uses Technorati.com to track Greek political blogs and records blogs with tags to the five Greek parliamentary parties during May 2009. Technorati.com is considered a reliable and popular blog search engine (Chin & Chignell, 2007; Kolari et al., 2006; Zafiropoulos & Vrana, 2008).

The five parliamentary Greek parties are: the New Democracy (ND) - the Christian Democratic party which was in government at the time when this manuscript was prepared, the Pan-Hellenic Socialist Movement (PASOK), which was the main opposition party, the Communist Party of Greece (KKE), the Coalition of the Left and Progress (SYRIZA) and the People's and the Orthodox's Rally, (LAOS) a right wing party that has newly entered the parliament and is mainly characterized by nationalist and populist practices and rhetoric.

Through the search, 101 blogs with «some authority» were found. According to Technorati.com, authority is the number of blogs linking to a website in the last six months. The higher the number, the more Technorati

Authority the blog has. In the authors view, considering blogs with some authority grants greater validity concerning blog selection because the analysis takes into account only the blogs which are active for a while and probably are considered reliable.

Next, the study analyzes incoming links between blogs through their blogrolls. Originating from the Social Networking Theory, this chapter presents the blogs' connectivity patterns using directed graphs, where blogs are presented as nodes and incoming links between blogs are presented as arrows. The chapter presents the hyperlinking patterns of the blogs using the Social Networking and graph theories. UCINET 6.0 for Windows is used to construct the network graph and to calculate graph theoretic indexes.

The directed graph, which presents the social network of blogs, is associated with its Adjacency Matrix. An Adjacency Matrix is a square non-symmetric binary data matrix where unity is placed in cell ij if blog i links blog j through the blogroll, else zero is placed in the cell. Let us consider a specific row of the Adjacency Matrix. This row is associated with a specific blog, say blog A, and unities in this row denote links from blog A to other blogs. By summing the unities of this specific row of the matrix, one can count the blogs that blog A links to. This sum of unities represents the number of outgoing links for blog A. On the other hand, every column of the adjacency matrix is associated with a specific blog, say blog B. By summing the unities of this specific column of the matrix, one can calculate how many blogs link to blog B. This sum of unities represents the number of incoming links for blog B. Outgoing and incoming links distribution, Centrality, Density, Betweenness, Components, Cliques, and Co-citations are the key graph theoretic indexes and properties that will be introduced and studied later on in this book as part of the analysis followed. These indexes and properties will help to figure out the degree of interconnectivity and cohesion of the specific network of blogs.

Besides using Social Networking Theory and graph theoretic properties, the study goes a step further by applying multivariate statistical analysis to the data of the adjacency matrix. This research adopts a method introduced by Zafiropoulos & Vrana (2008) for locating central blog groups in political blogging. The original idea is that political blogs are organized around central focal point blogs, where most of the informative conversation is taking place (Drezner & Farrell, 2004). Zafiropoulos & Vrana (2008) introduced a combination of Social Networking Theory, Multidimensional Scaling and Hierarchical Cluster Analysis to locate such groups by studying incoming links through blogrolls. By finding such groups, one can explore how bloggers are organized.

Multidimensional Scaling (MDS) is used in the analysis as a data reduction technique, on the one hand, and to quantify the original binary data on the other. The original data are those of the adjacency matrix which is a 101 by 101 binary matrix. Two problems arise when using the original data. First, there are too many variables which make further analysis practically impossible, and second, they are all binary. The second is not necessarily a problem, but usually continuous and quantitative variables permit us to use a wider variety of methods and techniques. MDS reproduces the original data and map them on a fewer dimensions space (namely two in this analysis) while the effort is to keep intact the distances among the original data and the newly reproduced data. "Stress" is a measure of goodness of fit between distances of original data and distances of the reproduced data. Better fit is assumed when stress is close to zero.

Next, Hierarchical Cluster Analysis (HCA) uses the quantified data from MDS as input, to produce clusters of blogs which have similar properties. Since cluster analysis is hierarchical, one needs some kind of decision rule in order to figure out how many clusters constitute a proper solution. Among several available tools and techniques to do this, scree plots are simple and efficient tools that can help researchers to decide, mainly through an inspection, on how many clusters to use. Scree-plots of number of clusters versus Wilks' Lambdas are useful for this purpose. Wilks' Lambda is a test statistic used in multivariate analysis of variance (MANOVA) and Discriminant Analysis to test whether there are differences between the means of identified groups (clusters in our case) of subjects on a combination of dependent variables (the two variables produces by MDS in our case). It ranges from zero to one and small values of Wilks' Lambda are associated with very good discrimination of the clusters. Thus the purpose of using a Wilks' Lambda scree-plot is to decide which the optimum number of clusters is. This optimum solution should conform to the following two properties: a) clusters in the solution produce a low value of Wilks' Lambda and b) a solution with more clusters will not decrease Wilks' Lambda significantly.

The original input variables are the columns of the adjacency matrix, thus, practically speaking, input variables are incoming links vectors and cases are the outgoing vectors for each blog. MDS quantifies the original data and produces two new coordinates for each original variable. This is done in such a way that original variables which were close together in the original data should be close together in the reproduced data. Variables which are close to each other in the reproduced data are associated with variables which were close to each other in the original data. In the original data, if two variables are close to each other, they probably present blogs with similar incoming links vectors. Thus, they present blogs which are linked by the same set of blogs. In this sense Cluster Analysis groups together blogs with similar incoming links vectors. Blogs in the same cluster are linked by nearly the same set of blogs. So in this way, the blogs in a formed cluster are regarded as having common characteristics or being a part of the same family, by way of the blogs which link them.

Some of the clusters that are produced by HCA gather the largest number of incoming links. If this happens, then they may serve as conversational focal points. This property might be associated with the skewed distribution of links, mentioned by Drezner & Farrell (2004) for political blogs: only few blogs have a very big number of incoming links while the rest, the majority of blogs, have only a small number of incoming links. This formation of focal point blogs is not necessarily a direct consequence of the skewness of incoming links distribution since clusters can be located in any circumstances and scaling of the data is not a prerequisite. This research also presents the distribution of incoming blog links. Statistical analysis was done using SPSS for Windows.

The study of the networks reveals how blogs are self-organized in groups around focal conversational points. Findings reveal whether the original hypothesis of Drezner & Farrell (2004) holds for the Greek case. Furthermore, by exploring inter-cluster linkages, the analysis explores whether Greek political blogs are characterized by a certain degree of polarization. The study of polarization may interpret whether blog clusters political affiliation is associated with blogs' interconnections.

Chapter 4

A SOCIAL NETWORKING THEORY APPROACH

SOCIAL NETWORKS INDEXES AND PROPERTIES

Links Distribution

In this chapter, the study follows a Social Networking Theory (SNT) approach. Incoming and outgoing links distributions and statistics give a sense of the size of blogs interconnectivity (Table 1). The number of outgoing links ranges from zero to 27 with a mean value of 4.87 and a standard deviation, which is greater than the mean, of (5.56). Dispersion of the number of incoming links is a little bit smaller since the maximum value is 16 and standard deviation is 3.46 while mean value is 4.87 as it is for incoming links.

Figures 4 and 5 show the graphic representation of the links distributions. Distributions appear skewed. Skewness is high for both distributions, however skewness of outgoing links is much higher that that of the incoming links. Skewness denotes the situation where only a few blogs have the biggest number of links, while the majority has only a few links. Therefore, from the analysis, it is obvious that this property holds for both incoming and outgoing links, but it is more intense for outgoing links. Skewness of links is a common property for several WEB2.0 applications. For example, on Flickr.com, a minority of users upload a large amount of pictures while the rest of the users upload only few (http://www.ted.com/talks/ clay_shirky_on_institutions_ versus_collaboration.html).

Table 1. Descriptive Statistics of the links

	N	Minimum	Maximum	Mean	Standard Deviation	Skewness
Incoming links	101	0	16	4.87	3.46	1.008
Outgoing links	101	0	27	4.87	5.57	1.806

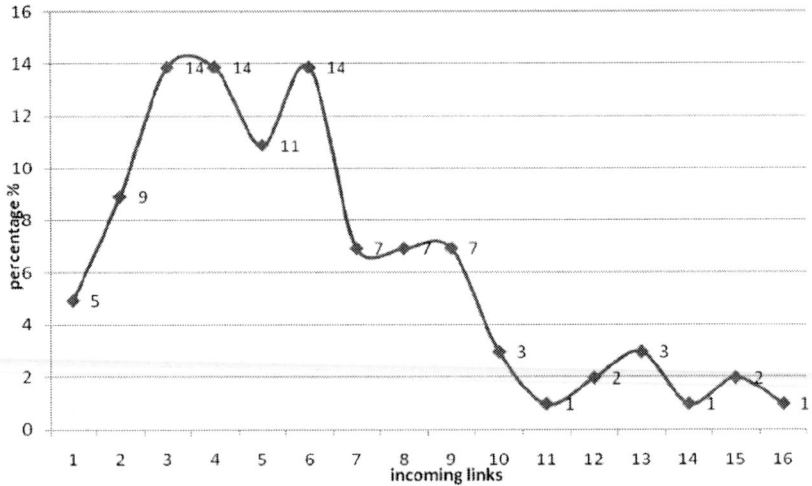

Figure4. A detailed presentation of the distribution of blogs incoming links.

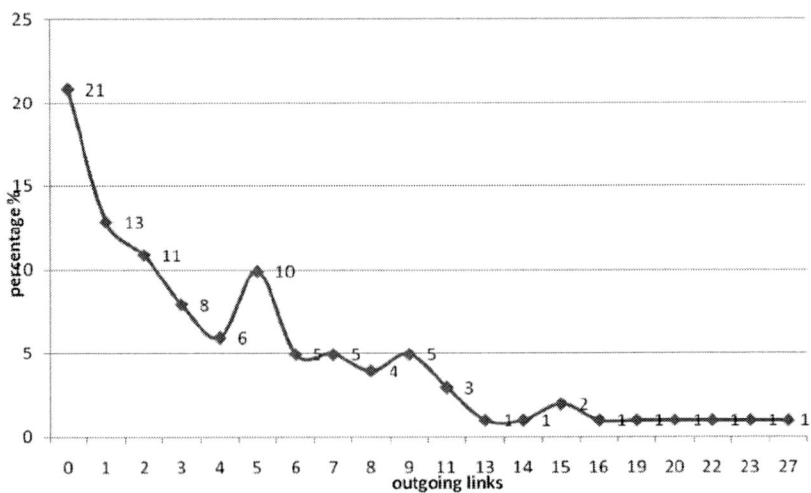

Figure 5. A detailed presentation of the distribution of blogs outgoing links.

CENTRALITY

The centrality of a node in a network is a measure of the structural importance of the node. A person's centrality in a social network affects the opportunities and constraints that they face. In this section, we consider two forms of centrality: density and betweenness.

DENSITY: Degree is simply the number of nodes that a given node is connected to. In general, the greater a person's degree, the more potential influence they have on the network, and vice-versa. For example, in a community network, a person who has more connections can spread information more quickly, and will also be more likely to hear more stuff. The greater a person's degree, the greater the chance that they will catch whatever is flowing through the network. The "density" of a binary network is the total number of ties divided by the total number of possible ties. In the case of Greek blog density the matrix average equals 0.0485 or 4.85% with a standard deviation 0.2149 or 21.49%. It becomes obvious that density is extremely low. On the other hand, the standard deviation is much larger than the mean. This implies that some blogs within the network have many links while many others have few to no links. The network as a whole seems lacking in interconnections, or there exists just few sparse links between some blogs.

BETWEENNESS: Loosely speaking, betweenness centrality is defined as the number of geodesic paths that pass through a node. It is the number of "times" that any node needs go through a given node to reach any other node by the shortest path. A node with high betweenness can serve as a liaison between disparate regions of the network. Betweenness is therefore a measure of the number of times a vertex occurs on a geodesic path. The normalized betweenness centrality is the betweenness divided by the maximum possible betweenness, expressed as a percentage. In this specific blogs' network mean betweenness equals 1.56 and its standard deviation equals 2.16. There is a lot of variation of normalized betweenness since standard deviation is much higher than the mean. This means that there exists some blogs that have the property of betweenness, while on the other hand there are blogs who do not have this property. Further analysis of the distribution of normalized betweenness for the 101 blogs reveals that 74% of them have the property of betweenness, while the remaining 26% of the blogs do not have this property.

The Network Centralization Index can be regarded as a measure for the network to have central points or areas with a greater number of paths than usual. The Network Centralization Index equals 9.14%. Despite a large variance of normalized betweenness, this index is low. In conclusion, the specific network has only few central blogs (concerning their links) and they differ in a medium degree from the rest regarding the property of centrality and the property of linking to others.

COMPONENTS AND CLIQUES

Cliques and components are sets of blogs that are connected in certain ways. Finding cliques is an attempt to understand whether blogs interact as linked friends in a social network.

COMPONENTS: A connected component is a maximal subgraph in which all nodes are reachable from every other. In a directed graph, two vertices are in the same weak component if there is a semi-path connecting them. In this directed network, two components were found. Regarding the components sizes, one component includes most of the blogs: 99% in one component and 1% in the other. It can be calculated that fragmentation, that is the proportion of blogs that cannot reach each other, equals 2%. This number is considered very small, and it probably means that in general, one user can navigate through these blogs using links from blogrolls, and practically with some effort and time spent; he/she could visit nearly all the blogs. However, things look a little different when studying cliques.

CLIQUES: For the needs of the analysis, the original adjacency matrix is symmetrised. This means that we consider that two blogs are connected if either one links to the other. A clique is a maximally complete subgraph. Maximal means that it is the largest possible subgraph: you could not find another node anywhere in the graph such that it could be added to the subgraph and all the nodes in the subgraph would still be connected. A complete graph is a simple graph in which every pair of distinct nodes-vertices is connected by an edge.

The number of cliques and the number of blogs who form cliques could be regarded as a measure of the volume of interconnectivity of the network. It shows how blogs are connected to form small companies of friends. 88% of the blogs form 193 cliques: 7 cliques (3.6%) with 5 blogs, 42 (21.8%) with 4 blogs, and 144 (21.8%) cliques with 3 blogs.

The line graph in Figure 6 presents how many blogs take part in several cliques at the same time. This is a measure of popularity and a measure of interconnection at the same time. Over 60% of the blogs in cliques are members of one to five cliques.

Concluding, we can say that group formation is high for Greek blogs. Links are few; however there is almost no isolated blogs, while groupings of blogs in small groups are often.

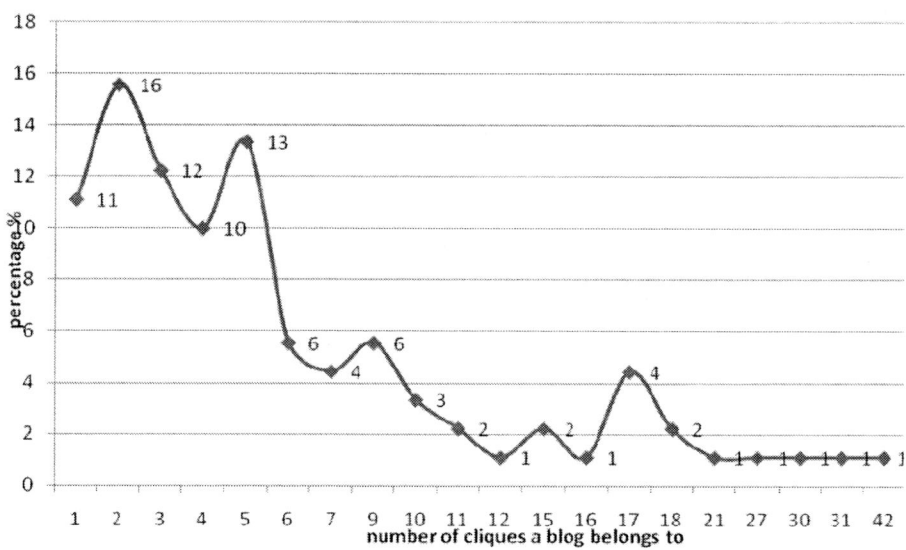

Figure 6. Distribution of blogs according to incorporation to cliques.

CO-CITATION

This section studies the way that blogs are co-cited by other blogs. That is- how they are linked simultaneously by more than one blog. Co-citations are used in the sense that they may be indexes of popularity of the blogs. Alternatively, they may be indexes of describing to what degree blogs recognize central – core groups of other blogs and in this way, they point to them. This is an important concept since it deals with the property of the network's self-realization. That is, the way that blogs have a view of what happens in the network, who links to whom, who are the most interesting blogs, etc. Let us revisit the example from the study of political blogging: within polarized political systems blogs are forming clusters around central blogs, which are considered reliable or have similar affiliations. Users of the Internet who wish to be informed quickly, locate the focal points of discussion and for economy of navigation, they read only the posts on these blogs. Bloggers also locate focal point blogs and place their posts along with a link to their blog. They thus expect that the readers of focal point blogs will also visit their blogs (Vrana & Zafiropoulos 2009, Zafiropoulos & Vrana 2008).

Co-citations are calculated by multiplying the transposed adjacency matrix by the original adjacency matrix. The product is a symmetric matrix. Its elements in cell ij is the number of blogs that link both blogs i and j. Therefore, for each blog a series of 101 numbers is produced: it presents the numbers of co-citations with any blog. Obviously, it is very hard to present all these findings. For economy, we use just a part of this data set. For each blog, the maximum co-citation with any other blog is chosen and is presented in Figure 7. For example, 5% of the blogs are not co-cited with any other blog, while 9% of the blogs are co-cited along with one other blog. More than 50% of the blogs are co-cited by 2, 3, 4 or 5 blogs through blogrolls. Co-citation within political blogs is low since one should consider that the reported numbers are the maximum co-citations for each blog.

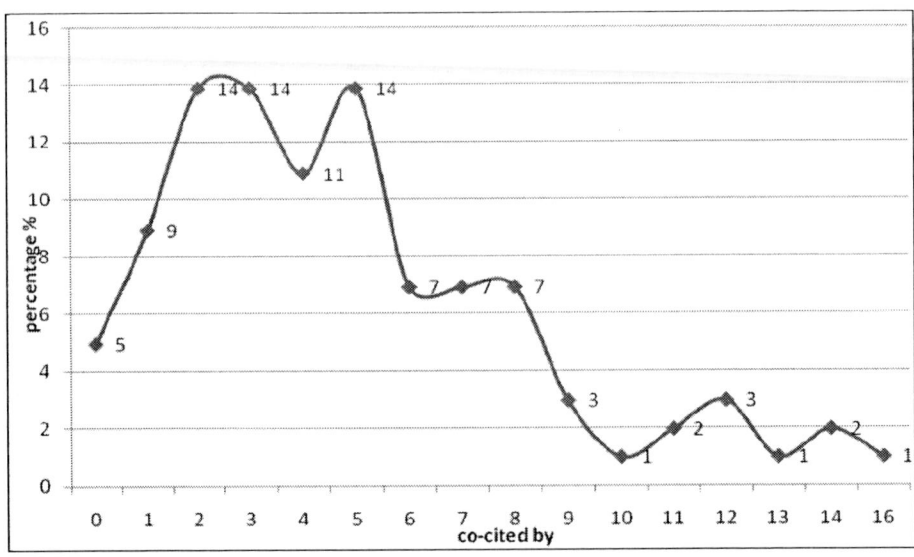

Figure 7. Distribution of blogs proportions according to co-citations.

Chapter 5

EXPLORING BLOGS COMMUNICATION PATTERNS

TESTING THE HYPOTHESIS OF CENTRAL BLOG GROUPS FORMATION

Drezner & Farrell (2004, p.13) mention "Blogs with large numbers of incoming links offer both a means of filtering interesting blog posts from less interesting ones, and a focal point at which bloggers with interesting posts, and potential readers of these posts can coordinate. When less prominent bloggers have an interesting piece of information or point of view that is relevant to a political controversy, they will usually post this on their own blogs. However, they will also often have an incentive to contact one of the large 'focal point' blogs, to publicize their post. The latter may post on the issue with a hyperlink back to the original blog, if the story or point of view is interesting enough, so that the originator of the piece of information receives more readers. In this manner, bloggers with fewer links function as "fire alarms" for focal point blogs, providing new information and links." Also they mention that: "We note that this implies that even while focal point blogs play a crucial mediating role, smaller blogs may sometimes have very substantial political impact by bringing information to the attention of focal blogs" (Drezner & Farrell, 2004; p.13). This chapter argues that "focal point" blogs are recognized as authority blogs by the blogger community and they may serve as the blogs' cores where the interesting and informational discussion is taking place. This property can be used to limit the analysis only to these blogs, excluding in this way other blogs which might be considered to have limited interconnections with other blogs or ones that are isolated. This property might be a consequence of the skewed distribution of links, also mentioned by Drezner & Farrell (2004).

To test whether this hypothesis holds for political blogging in Greece, this chapter examines the distribution of incoming links to the 101 blogs of the study. Figure 8 presents the histogram of blogs' incoming links and the scatter-plot of blog ranks according to incoming links (1 being the 1st blog in order, that is the blog linked the most) versus the percentage of incoming links. From these figures, it is obvious that there exists a slightly skewed distribution, which implies that few blogs gather the highest number of incoming links. This finding provides evidence that Drezner & Farrell's (2004) argument about the skewness of incoming links distribution holds true. Similar findings are produced when outgoing links are studied. Outgoing links are links from blogs' blogrolls to other blogs. They present the way that bloggers connect to other bloggers and they may reveal connectivity patterns between "neighbouring" bloggers. Skewness of outgoing links distribution is greater than that of incoming links, and this finding shows that only few bloggers are really engaged in connecting and pointing to others while the rest of (the majority) are linking to only a few or even no other blogs (Figure 9).

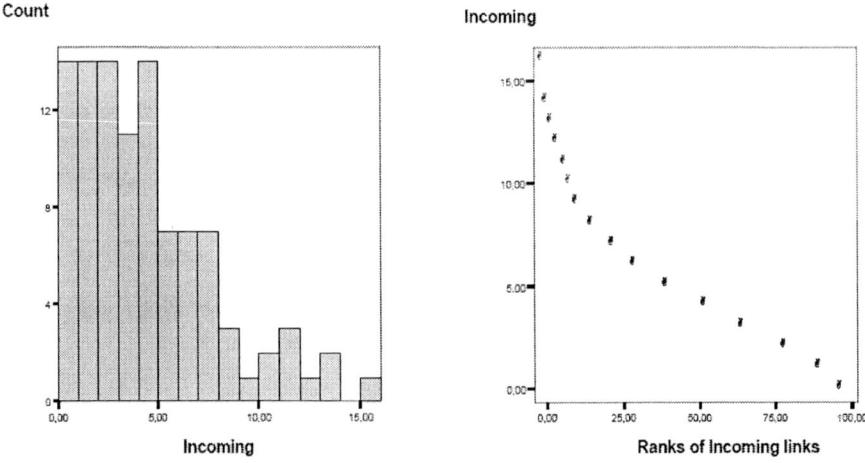

Skewness=1.008

Figure 8. Histogram of blogs' incoming links (left) and scaterplot of blogs' ranks according to incoming links vs actual number of incoming links (right).

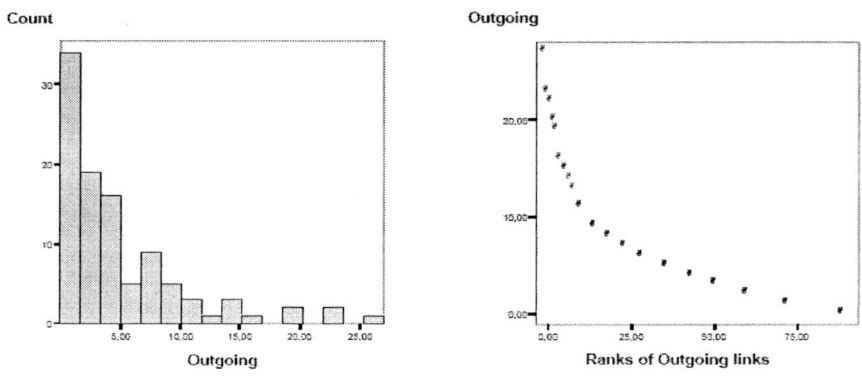

Skewness=1.806

Figure 9. Histogram of blogs' outgoing links (left) and scaterplot of blogs' ranks according to outgoing links vs actual number of outgoing links (right).

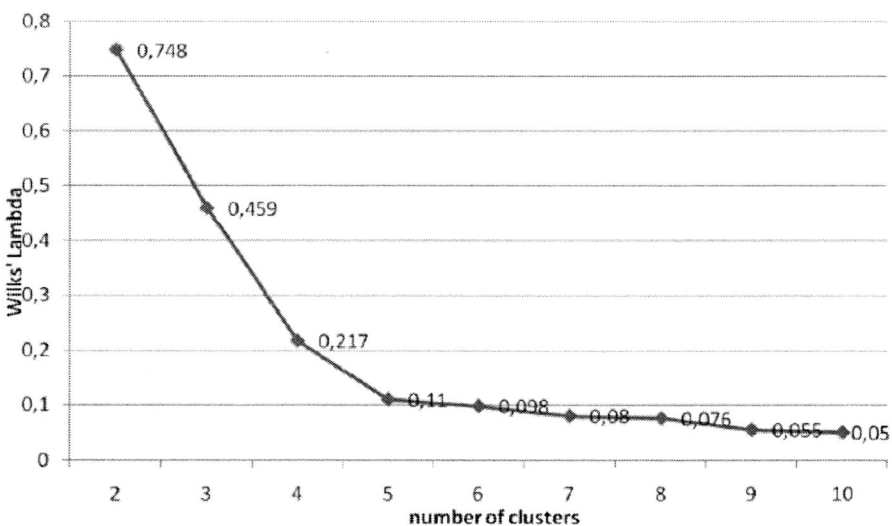

Figure 10. Scree plot of number of clusters vs Wilks' Lambdas.

To analyze this situation further and to locate core blog groups, statistical analysis using Multidimensional Scaling (MS presents a very good fit with Stress=0.07), followed by Hierarchical Cluster Analysis (HCA) is performed. Using a Wilks' Lambdas scree-plot it seems that the optimal number of blogs is five. Five clusters of blogs are formed regarding the incoming links (Figure

10). These clusters of blogs are described in Figure 11. When these clusters are studied in relation to their size and number of interconnections, some interesting results may arise.

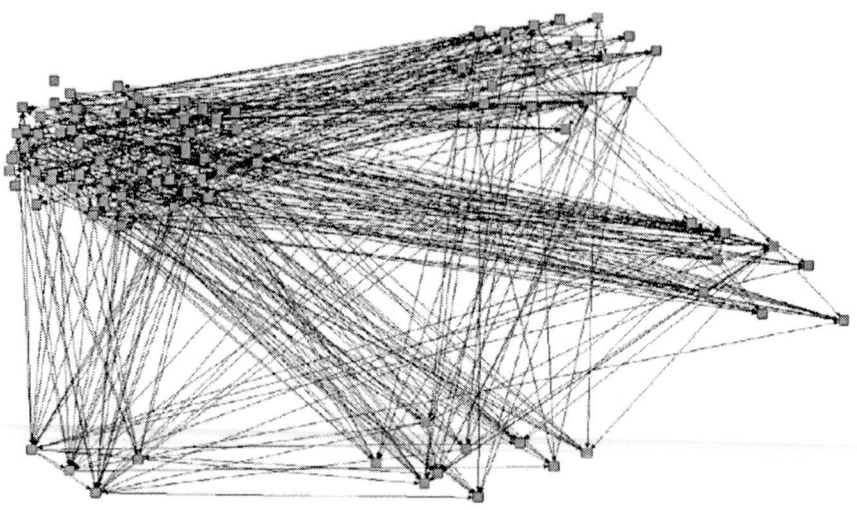

Figure 11. Blogs Social Network according to incoming links through blogrolls. Cluster 1 is placed at the upper left side of the graph, while clusters 2 to 5 are placed clockwise.

DESCRIPTION OF THE CLUSTERS

Statistics and Political - Ideological Affiliation

Table 2 presents the five clusters in descending order of size from the left to the right. Cluster 1, for example, contains more than 60% of the blogs. Although it is the most populous cluster, it presents the lowest rate of both incoming and outgoing links. Thus, the cluster contains the least active blogs regarding networking. Regarding the other clusters, there is strong negative correlation between cluster size and percentage of incoming links. The more we move towards clusters 5, the more the average incoming link rate is and the lower the size of the cluster becomes. This finding is summarized in Figure 12 where the negative correlation of the percentage of blogs in every cluster and the percentage of incoming links is obvious. Leaving out cluster 1, we could limit our analysis to and comment on the remaining four clusters, those that have many incoming links and small size at the same time. In particular, clusters 3, 4, and 5 in average, receive more than 9 incoming links. Because the recorded blogs are 101 (almost 100) this number can be directly interpreted as being a percentage, in the particular case of nearly 9%. Further, the most linked cluster consists of just four blogs and has an average of 10.7 of incoming links.

Regarding the tags of the blogs within each cluster, cluster 2 presents a high rate of tagging to the Communist Party of Greece (KKE) and SYRIZA, since 72.2% of the blogs within the clusters have tags to KKE while 50% of the blogs tag to SYRIZA. It is a group of blogs maintained mainly by leftists with references to KKE and SYRIZA. Cluster 3 tags mainly PASOK while it keeps tagging to the other parties in a very small quantity. Blogs in this cluster are maintained by members and friends of PASOK. Clusters 4 and 5 have high rates of tags to PASOK, ND and SYRIZA, while 50% of blogs in cluster 5 tag to KKE. These clusters are comprised of blogs, which provide information argumentation, speculation about society, and digital liberties (Table 3). It is very interesting to note that that the frequency of blogs with a Right wing affiliation is almost null (Table 3).

Table 2. Clusters properties

Clusters	1	2	3	4	5
Frequency	62	18	8	9	4
Percentage	61.4%	17.8%	7.9%	8.9%	4.0%
Outgoing links (average)	3.0	6.6	9.8	9.7	4.7
Incoming links (average)	2.8	6.2	9.3	9.5	10.7
Tags to PASOK	56.5%	44.4%	75.0%	66.7%	75.0%
Tags to ND	58.1%	44.4%	37.5%	77.8%	75.0%
Tags to SYRIZA	35.5%	50.0%	12.5%	77.8%	75.0%
Tags to KKE	32.3%	72.2%	25.0%	33.3%	50.0%
Tags to LAOS	24.2%	22.2%	12.5%	33.3%	25.0%

Table 3. Clusters' profiles and affiliation. Clusters sorted in ascending order of average number of incoming links

Cluster 2	Mainly Left, KKE and SYRIZA
Cluster 3	PASOK
Cluster 4	Left with broader speculation about society and democracy
Cluster 5	Left, digital liberties, information provision, discussion and argumentation

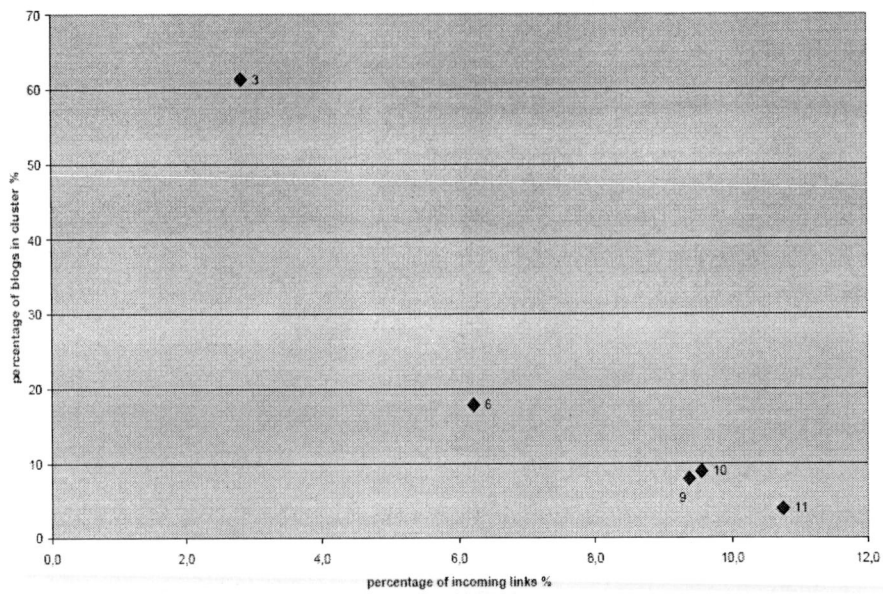

Figure 12. Scaterplot of clusters according to their characteristics: incoming links-number of blogs within clusters.

MEASURING POLARIZATION USING THE CLUSTERS INTERNAL COHERENCE AND EXTERNAL DIVERGENCE

Since clusters of blogs are, to a certain degree, affiliated with a political party or to some political orientation, it is important to explore 1) how strong the internal communication between blogs of the same cluster is. To answer this question we measure the number of incoming links among blogs within each cluster, 2) how strong the external divergence between clusters is. To answer this question we proceed by counting the incoming links among blogs of different clusters. In a polarized system, we expect intense communication among blogs of the same clusters or clusters of familiar affiliation on the one hand and weak communication among blogs of different clusters. Throughout this chapter, the number of links among blogs measures strength of communication. In this particular case polarization should be noticed when links within clusters are many while links among between clusters are few.

A cluster is considered partisan if all its blog members are of the same affiliation and they are interconnected to a high degree within the cluster. The particular political blogosphere is considered polarized if clusters are hardly interconnected while there is high enough interconnections within the clusters. In a polarized system there are only few incomings links from blogs of a cluster to the blogs of any other clusters, while there are a lot of links within each cluster.

Internal coherence of a cluster of blogs refers to the degree that the blogs within the cluster link to each other. In other words, when many blogs within the cluster connect to each other, internal coherence is high. Therefore, internal coherence maximum value equals the maximum number of incoming links within the cluster, and this equals n·(n-1) where n stands for the number of blogs within the cluster.

External divergence emerges when there are only few links between two clusters of blogs. If this happens then there are only sparse links from blogs of one cluster to blogs of the other. The maximum number of links between two clusters, equals n·m where n stands for the number of clusters for the one cluster and m the number of the clusters of the other.

To calculate indexes to describe coherence and divergence we work as follows:

At first, a 5 by 5 square matrix, matrix A, is formed that contains the maximum number of links between blogs, as described above. Thus in cell i,j of this matrix the maximum number of links that could appear joining cluster i with cluster j is calculated. Next, the actual number of links between cluster i and cluster j are calculated and are recorded in a matrix B. Cell i,j of this matrix contains the number of incoming links from cluster i to cluster j as they are recorded in the present research.

At the third step, cell i,j of matrix B is divided by cell i,j of matrix A and the outcome is presented as a percentage. This is an index of interconnection degree between clusters of blogs. The diagonal elements of this final product matrix should be high in order for the clusters to have high internal coherence, while the rest of the cells should have small values in order for the clusters to have a low degree of connectivity i.e. a high degree of divergence.

Table 4 contains the outcomes of these calculations. The number in any cell i,j denotes the percentage of incoming links to cluster j from cluster i. Cluster 1 appears to be the least active regarding connections. Blogs in cluster 1 neither receive nor do they send any links to blogs of the other clusters. Furthermore, internal coherence for this cluster is very low, since only 2% of all the available links are really used. For the rest of clusters internal coherence is quite high. It nearly equals 18% except for cluster 3 (affiliated to PASOK) where internal coherence rises to 34%- being double that of the internal coherence of each one of the clusters 2, 4, and 5. Having a very clear political affiliation to Pan-Hellenic Socialist Movement and a high degree of internal coherence, cluster 3 seems to be the most partisan cluster. It consists of very active members who communicate and connect to each other. In addition, cluster 3 has high links percentages to and from clusters 4 and 5.

Cluster 2 has a fair degree of internal coherence (18%) and very low degrees of external connectivity to and from the other clusters. However, links percentages to clusters 4 and 5 both equal 11% and incoming links percentage from cluster 4 to cluster 2 equals 12%. It is a more isolated cluster of blogs with a fair internal coherence. Taking into consideration that this cluster contains blogs affiliated mostly to KKE and SYRIZA, we could define it as partisan with few ties to other clusters.

Cluster 4 presents a good degree of internal coherence, 18%, and good ties (links percentages) to clusters 2, 3, and 5. Its links percentages to and from cluster 5 are very high, 25% and 22% respectively. Clusters 4 and 5 are really well connected. This property might be associated with the fact that they share common beliefs and ideas.

Cluster 5 looks very similar to cluster 4, regarding percentages of incoming links. However, it differs from cluster 4 in the sense that while cluster 2 links fairly well to it, cluster 5 does not link to cluster 2 (1%). Blogs in cluster 2 might regard blogs of cluster 5 as interesting or being politically close to them, but the opposite does not seem to hold.

Table 4. Percentages of incoming links with respect to the maximum number of incoming links between clusters

	Cluster 1	Cluster 2	Cluster 3	Cluster 4	Cluster 5
Cluster 1	2%	3%	7%	6%	8%
Cluster 2	3%	18%	4%	11%	11%
Cluster 3	8%	2%	34%	18%	16%
Cluster 4	7%	12%	13%	18%	25%
Cluster 5	2%	1%	13%	22%	17%

In conclusion, clusters 4 and 5 present the highest degrees of interconnections or ties. They both are fairly well connected to cluster 3, but compared to it they have half of its internal coherence. Cluster 3 is the cluster with the most active and interconnected members. Clusters 3, 4 and 5 form a larger extended cluster of smaller connected clusters. Cluster 2 has more sparse connections to clusters 3, 4, and 5 and seems to be isolated from the other clusters while still containing active members. Figure 13 presents the strength of interconnections between different clusters of blogs. Larger arrowheads are used to present higher percentages of incoming links. The block of clusters 3, 4, and 5 is presented at the lower part of the graph. The connections between these clusters are denoted with larger arrowheads.

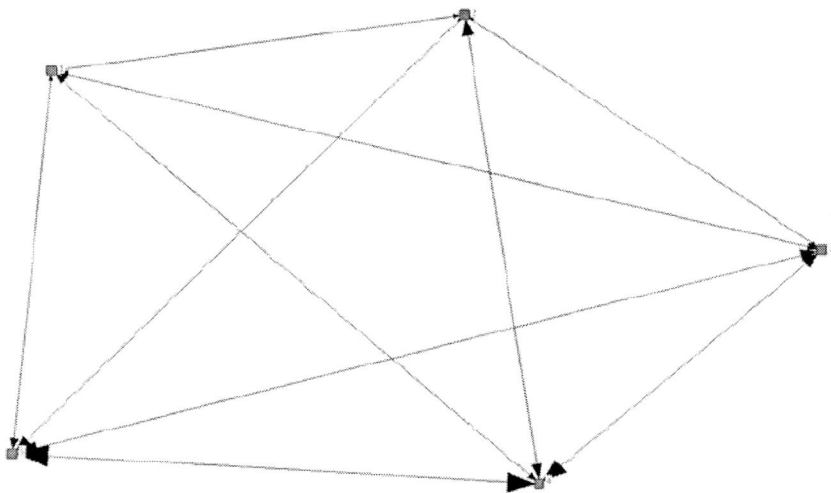

Figure 13. Strength of interconnections between clusters of blogs.

Conclusion

The book presented a broad hyperlink analysis of Greek political blogs. Hyperlinks are important to bloggers not only to maintain some communication with each other, but to determine their digital territory as well. Regarding the social networking analysis of the blogs, hyperlinks are few, yet connectivity is fairly high. This means that in general, there is always a path (through blogs hyperlinks) that can lead a user from a blog to another blog, simply by following consecutive links. In this sense, isolated blogs hardly exist. In addition, co-citation exists but it is relatively low. Blogs are linked by groups of other blogs. This supports the hypothesis that bloggers, to a small degree, manage to locate central blogs groups and they gather around them, linking to them. They form, in this way, groups of blogs around central and significant clusters of blogs, possibly because they find that these central clusters of blogs are of the same affiliation or they are reliable or informative.

A portion of bloggers is self-organized in central clusters of blogs having the same political or ideological affiliation, and to a certain degree, they develop interconnections within each cluster. The degree of interconnections varies across clusters. Some clusters present high degrees of internal coherence, but in any case, internal coherence for every such cluster is medium. In addition, connection between clusters exists. In this way, this system of clusters of blogs reveals that blogs, most of them having a left-wing orientation, are organized in a system, which is partisan but not very polarized. Furthermore, when looking at the big picture, it becomes clear that the basic hypothesis of focal conversational points creation, supported in the literature, is valid to a certain degree for the Greek case too. Blogs are organized in groups around clusters of central blogs, which share a common affiliation. These core groups can be located using multivariate statistical techniques presented in this chapter. Having these properties, the Greek political blogosphere, although limited, seems to have grown from the early stage phase to the phase where self-organization and self-realization have arisen.

REFERENCES

Adamic, L.& Glance, N. (2005). The Political Blogosphere and the 2004 U.S. Election: Divided They Blog. In proceedings of the WWW-2005, workshop on the Weblogging Ecosystem. http://www.blogpulse.com/ papers/ 2005/ AdamicGlanceBlogWWW.pdf

Albrecht, S., Lübcke, M. & Hartig-Perschke, R. (2007). Weblog Campaigning in the German Bundestag Election 2005. *Social Science Computer Review*, 25(4), 504-520. Allan, S. (2002). Reweaving the Internet: Online news of September 11. In B. Zelizer & S. Allan (Eds.), *Journalism after September 11* (pp. 119-140). New York: Routledge.

Anderson, P. (2007). What is Web 2.0? Ideas, technologies and implications for education. *JISC Technology and Standards Wa*tch, Feb. 2007

Andrews P. (2003). Is blogging journalism? *Nieman Reports*, 57(3), 63-64. http://download.101com.com/syllabus/conf/summer2004/PDFs/w01.pdf

Augar, N., Raitman, R. & Zhou, W. Teaching and learning online with wikis. In R. Atkinson, C. McBeath, D. Jonas-Dwyer & R. Phillips (Eds), Beyond the comfort zone: Proceedings of the 21st ASCILITE Conference (pp. 95-104). Perth, 5-8 December. http://www.ascilite.org.au/conferences/ perth04/procs/augar.html

Balancieri, R., Cuel, R., & dos Santos Pacheco, R.C. (2007). Social Network Analysis for Innovation and Coordination. Proceedings of I-KNOW '07 Graz, Austria, September 5-7, 2007, http://triple-i.tugraz.at/blog/wp-content/uploads /2008/ 11/9_social-network-analysis-for-innovation-and-coordiniation.pdf

Baker, S. & Green, H.(2005). Blogs will change your business. *Business Week*, 5 February , 3931, 56-65.

Banning, S., & Trammell, K. D. (2006). Revisiting the issue of blog credibility: A national survey. Paper presented at the annual conference at the Association for Education in Journalism & Mass Communication, August, San Francisco, CA

Barabak, M. (2006). Campaign '08 Preview: Podcasting Politicians. Los Angeles Times. http://articles.latimes.com/2006/jul/21/nation/na-fragment 21

Barger, J. (1997). FAQ: Weblog Resources. http://www.robotwisdom.com/weblogs/

Bar-Ilan, J. (2007). The use of Weblogs (blogs) by librarians and libraries to disseminate information. *Information Research*, 12(4), paper 323. http://informationr.net/ir/12-4/paper323.html (6/4/2007)

BBC News (2006) Egypt arrests another blog critic, Monday, 20 November 2006. http://news.bbc.co.uk/2/hi/middle_east/6164798.stm

Best, D (2006). *Web 2.0 Next Big Thing or Next Big Internet Bubble?* Lecture Web Information Systems. Technische Universiteit Eindhoven

Betts, D. J., & Glogoff, S. J. (2004). Instructional models for using weblogs in elearning: A case study from a virtual and hybrid course. *Syllabus 2004*. http://chnm.gmu.edu/digitalhistory/links/pdf/chapter1/1.41.pdf

Bielenberg, K. and Zacher, M. (2006) Groups in Social Software: Utilizing Tagging to Integrate Individual Contexts for Social Navigation, Masters Thesis submitted to the Program of Digital Media, Universität Bremen. http://bielenberg.info/thesis.pdf

Bloom, J. (2003). The blogosphere: how a once-humble medium came to drive elite media discourse and influence public policy and elections. Paper presented to the American Political Science Association, Philadelphia, PA, August

Bloom, J. & Kerbel, M. (2005). The blogosphere and the 2004 elections. Paper presented at the American Political Science Association Annual Conference, Washington, DC, September.

Blood, R. (2002). *The Weblog Handbook: Practical Advice on Creating and Maintaining Your Blog.* Cambridge MA: Perseus Publishing.

Blood, R. (2003). Weblogs and Journalism: Do They Connect? *Nieman Reports*, 57(3), 61-63. http://download.101com.com/syllabus/conf/summer2004/PDFs/w01.pdf

Boyd, D. (2006). A Blogger's Blog: Exploring the Definition of a Medium. *Reconstruction* 6(4). http://reconstruction.eserver.org/064/boyd.shtml

Boyd, D. M., & Ellison, N. B. (2007). Social network sites: Definition, history, and scholarship. *Journal of Computer-Mediated Communication, 13*(1), article 11. http://jcmc.indiana.edu/vol13/issue1/boyd.ellison.html

Brady, M. (2005). Blogging: personal participation in public knowledge building on the web Chimera Working Paper Number: 2005, 02. www.essex.ac.uk /chimera/content/pubs/wps/cwp-2005-02-blogging-in-the-knowledge- society-mb.pdf

Chadwick, A.(2008). Web 2.0: New Challenges for the Study of E-Democracy in an Era of Informational Exuberance. *I/S A Journal of Law and Policy for the Information Society,* 4 (3),10-42

Chau, M. & Xu, J. (2008) Using Web Mining and Social Network Analysis to Study The Emergence of Cyber Communities In Blogs. *Integrated Series in Information Systems,* 18, 1571-0270

Chin, A. & Chignell, M. (2007). Identifying communities in blogs: roles for social network analysis and survey instruments. *Int. J. Web Based Communities*, 3(3), 345–363.

Connor, A. (2005), Not just critics, BBC News, http://news.bbc.co.uk/2 /hi/ uk_news/magazine/4111330.stm

Cornfield, M., Carson, J., Kalis, A. & Simon, E. (2005). Buzz, blogs and beyond. Pew Internet & American Life Project. www.pewinternet.org/ppt/buzz_blogs_beyond_ final05-16-05.pdf

Drezner, D. & Farrell, H. (2004). The power and politics of blogs, paper presented at the Annual Meeting of the American Political Science Association, Washington, DC, August.http://www.utsc.utoronto.ca/~farrell/ blogpaperfinal.pdf

Du, H. & Wagner, C.(2006). Weblog success: Exploring the role of technology. International. *Journal of Human-Computer Studies,* 64, 789-798.

Duffy, P. and Bruns, A. (2006) The Use of Blogs, Wikis and RSS in Education: A Conversation of Possibilities. In Proceedings *Online Learning and Teaching Conference* 2006, pages pp. 31-38, Brisbane. http://eprints.qut.edu.au

Dwivedi, M., Shibu. T.P., & Venkatesh, U. (2007). Social software practices on the internet. Implications for the hotel industry. *International Journal of Contemporary Hospitality Management*, 19(5), 415-426.

Edelman (2009). Social Pulpit. The Barack Obama's Social Media Toolkit http://www.edelman.com/image/insights/content/Social%20Pulpit%20-%20Barack %20Obamas%20Social%20Media%20Toolkit%201.09.pdf

Efimova, L. & Hendrick, S. (2005). In search for a virtual settlement: an exploration of Weblog community boundaries. https://doc.novay.nl/dsweb/Get/Document-46041

Efimova, L., Hendrick, S. & Anjewierden, A. (2005). Finding "the life between buildings": an approach for defining a weblog community. AOIR Internet Research 6.0: Internet Generations, Chicago. https://doc.novay.nl/dsweb/Get/Document-55092/AOIR_blog_communities.pdf

Farmer, J. (2004). Communication dynamics: Discussion boards, Weblogs and the development of communities of inquiry in online learning environments. In R. Atkinson, C. McBeath, D. Jonas-Dwyer, & R. Phillips (Eds.), *Beyond the comfort zone (pp. 274-283). Proceedings of the 21st ASCILITE Conference.*

Farrell, H., & Drezner, D. W. (2008). The power and politics of blogs. *Public Choice*, 134, 15–30.

Festa, P. (2003). Blogging comes to Harvard. Newsmaker. http://news.com.com/2008-1082-985714.html?tag=fd_nc_1

Fuchs, C. (2007) Towards a dynamic theory of virtual communities. *International Journal of Knowledge and Learning*, 3(4/5), 372 - 403

Garrett, J. (2004). User experience analysis: Presidential campaign sites. Report prepared for Adaptive Path.

Gilchrist, A. (2007). Can Web 2.0 be Used Effectively Inside Organisations? *Bilgi Dünyası* 8(1), 123–139

Gil de Zúñiga, H., Veenstra, A., Vraga, E., & Shah, D. (2010) Digital Democracy: Reimagining Pathways to Political Participation. *Journal of Information Technology & Politics*, 7(1), 36 - 51

Gill, K.(2004) How can we measure the influence of the blogosphere? *WWW2004*, May 17–22, 2004, New York, NY USA. http://faculty.washington.edu/kegill/ pub/ gill_blogosphere_www2004.pdf

Glover, D. (2006). The Rise of Blogs. National Journal, January 21, 2006, republished online at Daniel Glover's Beltway Blogroll blog http://beltwayblogroll.nationaljournal.com/archives/2006/01/the_rise_of_blo.php

Golder, S. and Huberman, B. A. (2005). The Structure of Collaborative Tagging Systems. Technical report, Information Dynamics Lab, HP Labs.

Godin, S. (2006). Flipping the funnel. Give Your Fans the Power to Speak Up. http://www.duoconsulting.com/sites/default/files/resources/FlippingFunnelPRO.pdf

Graf, J. (2006). The Audience for Political Blogs. New research on Blog Readership.GW's Institute for Politics, Democracy & the Internet. George Washington University, Washington, DC http://www.ipdi.org/ uploadedfiles/ audience %20for%20political%20blogs.pdf

Graf , J. & Darr, C. (2004). Political Influentials Online in the 2004 Presidential Campaign, Institute for Politics, Democracy and the Internet, George Washington University, Washington, DC, February, http://www.ipdi.org/UploadedFiles/ political%20influentials .pdf

Grieve, J., Biber, D., Friginal, E. and Nekrasova,T. (2010).Variation Among Blogs: A Multi-dimensional Analysis. In Mehler, Sharoff. Rehm and Santini (eds.) Genres on the Web: Corpus Studies and Computational Models. New York: Springer-Verlag.

Gumbrecht, M. (2004). Blogs as "protected space" WWW 2004. Workshop on the Weblogging Ecosystem: Aggregation, Analysis and Dynamics.

Guo, X., Vogel, D. Zhou, Z., Zhang, X. & Chen, H. (2008). Chaos Theory as a Model for Interpreting Weblog Traffic. Proceedings of the 41st Hawaii International Conference on System Sciences - 2008

Hepburn, C. (2007). web 2.0 for the tourism and travel industry. http://www.ebusinessforum.gr/engine/index.php?op=modload&modname =Downloads&action=downloadsviewfile&ctn...el

Herring, S. C., Scheidt, L. A., Bonus, S. & Wright, E. (2004). Bridging the gap: A genre analysis of weblogs. Proceedings 37th Annual HICSS Conference 2004. Big Island, Hawaii

Herring,C., Kouper, I.Paolillo, J., Scheidt, L-A.,Tyworth,M., Welsch, P., Wright, E., & Yu, N. (2005). Conversations in the Blogosphere: An Analysis "From the Bottom Up" Proceedings of the Thirty-Eighth Hawai'i International Conference on System Sciences (HICSS-38). Los Alamitos

Hiler, J. (2002). Blogs as disruptive tech: How weblogs are flying under the radar of the content management giants. http://www.webcrimson.com/ ourstories/ blogsdisruptivetech.htm

Hoogenboom, T., Kloos, M., Bouman, W. & Jansen, R. (2007). Sociality and learning in social software. *International Journal of Knowledge and Learning*, 3(4/5), 501 - 514.

Hsu, C-L., & Lin, C-C. (2008). Acceptance of blog usage: The roles of technology acceptance, social influence and knowledge sharing motivation. *Information and Management,* 45, 65-74.

Jackson, N. (2006). Dipping their big toe into the blogoshpere. The use of weblogs by the political parties in the 2005 general election. *Aslib Proceedings: New Information Perspectives*, 58(4), 292-303

Johnson, T. & Kaye, B. (2004). Wag the blog: how reliance on traditional media and the internet influence credibility perceptions of weblogs among web users. *J&MC Quarterly*, 81(3), 622-642.

Johnson, T. J., & Kaye, B. K. (2007). Choosing is believing? How Web gratifications and reliance affect Internet credibility among politically interested users. Paper presented to the Association for Education in Journalism and Mass Communication, August, Washington, D.C.

Johnson, T. J., Kaye, B. K., Bichard, S. L., & Wong, w. J. (2007). Every blog has its day: Politically-interested Internet users' perceptions of blog credibility. *Journal of Computer-Mediated Communication*, 13(1), paper 6. http://jcmc.indiana.edu/vol13/issue1/johnson.html

Jüch, C. & Stobbe, A (2005). Blogs: The new magic formula for corporate communications? Economics. Digital economy and structural change. *Deutsche Bank Research*, 53, 22 August.

Kahn, R., & Kellner., D.(2004). New media and Internet activism: From the "Battle of Seattle" to blogging. *New Media & Society*,6(1), 87-95.

Karampasis, Z. (2007).Greek-speaking bloggers, http://ereuna. wordpress .com/ [In Greek]

Kelleher, T., and Miller, B. M. (2006). Organizational blogs and the human voice: Relational strategies and relational outcomes. *Journal of Computer-Mediated Communication, 11*(2), article 1. http://jcmc.indiana.edu/vol11/issue2/kelleher.html

Kenix, L.J. (2009). Blogs as Alternative. *Journal of Computer-Mediated Communication*, 14, 770-822.

Kerbel, M. R., & Bloom, J. D. (2005). Blog for America and civic involvement. *The Harvard International Journal of Press/Politics*, 10, 3-27.

Kim, H-N. (2008). The phenomenon of blogs and theoretical model of blog use in educational contexts. *Computers & Education*, 5, 1342–1352.

Koh, A., Lim,A., Soon,Mg-Ee., Detenber, B. & Cenite, M. (2005). Ethics in Blogging. Singapore Internet Research Centre. http://www.ntu.edu.sg/sci/sirc/

Kolari, P., Finin, T. & Joshi, A. (2006) SVMs for the Blogosphere: Blog Identification and Splog Detection. In: AAAI Spring Symposium on Computational Approaches to Analysing Weblogs. Maryland: University of Maryland.

Kolbitsch, J. & Mauer, H. (2007). The Growing Importance of e-Communities. *Web Lecture Notes in Business Information Processing,* 1,19-37

Kosonen, M., Henttonen, K., & Ellonen, H-K.(2007). Weblogs and internal communication in a corporate environment: a case from the ICT industry. *International Journal of Knowledge and Learning*, 3(4/5), 437-449

Krackhardt, D. (1996). Social Networks and the Liability of Newness for Managers. *Trends in Organizational Behavior*. In Cooper, C.L. and Rousseau D. M. (eds.), (pp. 159-173). John Wiley and Sons, Ltd. New York, NY.

Krishnamurthy, S. (2002). The Multidimensionality of Blog Conversations: The Virtual Enactment of September 11. In Maastricht, The Netherlands: Internet Research 3.0

Lankshear, C. & Knobel M. (2003). Do-It-Yourself Broadcasting: Writing Weblogs in a Knowledge Society Paper presented to the American Education Research Association Annual Meeting. Chicago, April 21 http:// www.ballarat.edu.au

Lasica, J. D. (2001). Blogging as a form of journalism. USC Annenberg. *Online Journalism Review*. http://www.ojr.org/ojr/workplace/1017958873.php

Lasica, J. D. (2002). Weblogs: A new source of news. In R. Blood (Ed.), *We've Got Blog: How Weblogs are Changing Our Culture* (pp. 171-182). Cambridge, MA: Perseus Publishing

Lilleker, D. & Jackson, N. (2008). Politicians and Web 2.0: the current bandwagon or changing the mindset? Paper presented at Web 2.0: an International Conference, New Political Communication Unit, Department of Politics and International, Relations, Royal Holloway, University of London, April 17-18, 2008.

Lu, H-P. & Hsiao, K-L. (2007). Understanding intention to continuously share information on weblogs. *Internet Research*, 17(4), 345-361.

Maag, M. (2005). The potential use of "Blogs" in nursing education. *CIN: Computers, Informatics, Nursing*, 23, 16–24.

MacKinnon, R. (2007). Blogs and China Correspondence: How foreign correspondents covering China use blogs. A paper presented at *The World Journalism Education Congress* (WJEC) Singapore, June 25-28, 2007

Makice, K. (2006). PoliticWiki: exploring communal politics. International Symposium on Wikis. *Proceedings of the 2006 international symposium on Wikis*, 105 - 118

Marin, A. & Wellman, B. (2010). Social Network Analysis: An Introduction, in *Handbook of Social Network Analysis*. Edited by Peter Carrington and John Scott. London: Sage.

Marlow, C. (2004). Audience, structure and authority in the weblog community. The 54th Annual Conference of the International Communication Association, 2004. http://www.researchmethods.org/ICA2004.pdf .

Martino, F. & Spoto, A. (2006). Social Network Analysis: A brief theoretical review and further perspectives in the study of Information Technology. *PsychNology Journal*, 4(1), 53 – 86.

McKenna, L & Pole, A. (2003). Do Blogs Matter? Weblogs in American Politics American Political Science Association, Annual Meeting Chicago, IL, September 2004

Merritt, D. B. (1998). *Public journalism and public life: Why telling the news is not enough.* Mahwah, NJ: Lawrence Erlbaum Associates

Mishne, G. & Glance, N. (2006). Leave a reply: An analysis of weblog comments. WWW 2006 May 22–26, 2006, Edinburgh, UK. http://www.blogpulse.com/ www2006-workshop/papers/wwe2006-blogcomments.pdf

Nardi, B. Schiano, D. & Gumbrecht, M (2004) 'Blogging as social activity, or, would you let 900 million people read your diary?', Proceedings of the 2004 ACM Conference on Computer Supported Cooperative Work, New York, USA, 222–231.

O'Reilly, T. (2005) http://www.oreillynet.com/pub/a/oreilly/tim/news/2005/09/30/what-is-web-20.html

O'Reilly, T. (2006). Web 2.0 compact definition: trying again, http://radar.oreilly.com/archives/2006/12/web_20_compact.html.

Osimo D.(2008) Web 2.0 in Government: Why and How. European Commission. Joint Research Centre. IPTS. *Technical Report EUR 23358 EN (2008)*.

Park, H-W. & Jankowski, N. (2008). A hyperlink network analysis of citizen blogs in South Korean Politics. *Javnost-the Public*, 15(2), 57-74.

Pedley, P. (2005). International phenomenon? Amateur journalism? Legal minefield? Why information professionals cannot afford to ignore weblogs. *Business Information Review*, 22(2), 95-100.

Pohlig, C. (2003) How a Newspaper Becomes H.I.P. *Neiman Reports* (Winter): 24–6. http://www.nieman.harvard.edu/reportsitem. aspx?id=100906

Punie, Y. & Cabrera M. (2005). The Future of ICT and Learning in the Knowledge Society. Report on a Joint DG JRC-DG EAC Workshop Seville, 20-21 October 2005

Robinson, S (2006). The mission of the j-blog: Recapturing journalistic authority online. *Journalism*, 7(1), 65–83.

Rosenbloom, A.(2004). The Blogosphere. *Commun ACM*, 47(12), 31-33.

Regan T. (2003). Weblogs threaten and inform traditional journalism. *Nieman Reports* (Fall) ,57(3), 68–70.

Rutigliano, L. (2007). Emergent communication networks as civic journalism. In M. Tremayne (Ed.), *Blogging, citizenship and the future of media* (pp. 225-237). New York: Routledge.

Sigala M. (2008). Developing and implementing an eCRM 2.0 strategy: usage and readiness of Greek tourism firms. In O'Connor,P. Höpken, W. and Gretzel, U. (Eds) Information and Communication Technologies in Tourism 2008, (pp. 463-474.) Springer Verlag: Wien.

Sigala, M. (2009), "WEB 2.0, social marketing strategies and distribution channels for city destinations: enhancing the participatory role of travelers and exploiting their collective intelligence", In Gascó-Hernández, M and Torres-Coronas, T. (Eds.) *Information Communication Technologies and City Marketing: Digital Opportunities for Cities around the World*, IDEA Publishing, pp. 220 – 244

Sroca, N. (2007). Understanding the political influence of blogs. Institute for Politics, Democracy, & the Internet, Washington, D.C. www.ipdi.org/UploadedFiles/ PoliticalInfluenceofBlogs.pdf

Su, N-M, Wang, Y. & Mark G. (2005). Politics as Usual in the Blogosphere. Proceedings of the 4th International Workshop on Social Intelligence Design

Technorati (2008). State of the blogosphere 2008. http://technorati.com/ blogging/state-of-the-blogosphere/who-are-the-bloggers/TechWatch (2007). What is web 2.0. www.jisc.ac.uk/techwatch

Thaeler, J. (2008). 75% of Journalists Get Story Ideas from Blogs. http://www.marketingpilgrim.com/2008/01/75-of-journalists-get-story-ideas-from-blogs.html

Thompson, G. (2003). Weblogs, warblogs, the public sphere, and bubbles. Transactions, 2003. 7(2). http://transformations.cqu.edu.au /journal /issue_07 /paper_02.shtml

Trammell, K. & Keshelashvili, A. (2005). Examining the new influencers: A self-presentation study of A-list Blogs. *Journalism and Mass-Communication Quarterly*, 82(4), 968-982.

Trammell K., Williams A., Postelnicu M. & Landreville K. (2006). Evolution of Online Campaigning: Increasing Interactivity in Candidate Web Sites and Blogs Through Text and Technical Features. *Mass Communication and Society*, 9(1), 21-44.

Valenzuela, S., Park, N. & Kee, K. (2008). Lessons from Facebook: The Effect of Social Network Sites on College Students' Social Capital. 9th International Symposium on Online Journalism, Austin, Texas, April 4-5, 2008

Vrana, V. & Zafiropoulos, K. (2009). A social network analysis of travel blogs. In Proceedings of MCIS 2009. Information Society Research, Education, Policy and Practice in the Mediterranean Region. Athens, September 25-27.

Walker, J. (2007). Weblog, Definition for the Routledge Encyclopedia of Narrative Theory.

Wall, M. (2004). Blogs as Black Market Journalism: A New Paradigm for News. Berglund Center for Internet Studies, http://bcis.pacificu.edu/journal /2004/ 02/ wall.php

Wallsten, K. (2005). Political Blogs and the Bloggers Who Blog Them: Is the Political Blogosphere and Echo Chamber? Paper Presented at the American Political Science Association Annual Meeting Washington, D.C. September 1-4, 2005. http://www.journalism.wisc. edu/ blog-club/Site/Wallsten.pdf

Weil, D. (2003). Top 20 Definitions of Blogging. http://www.marketing profs.com /3/weil9.asp

Williams, J. B., & Jacobs, J.(2004). Exploring the use of blogs as learning spaces in the higher education sector. *Australasian Journal of Educational Technology*, 20, 232–247.

Williams , C. & Gulati, G. (2007). Social Networks in Political Campaigns: Facebook and the 2006 Midterm Elections. Annual Meeting of the American Political Science Association Chicago, Illinois, August 30 – September 2, 2007

Williams, A.P., Trammell K., Postelnicu M., Landreville, K. and Martin, J. (2005). *Journalism Studies*, 6(2), 177-186.

Yap, R., Muirhead, B. and Keefer, J. (2005). Blog RUBRIC: Designing your Business Blog. *International Journal of Instructional Technology and Distance Learning*, 2(11), 53-60

Zafiropoulos, K. & Vrana, V. (2008). A Social Networking Exploration of Political Blogging in Greece. Official Proceedings of 1st World Summit on the Knowledge Society, Lytras M. D., Carroll J. M., Damiani E., and Tennyson R. D. (Eds). Emerging Technologies and Information Systems for the Knowledge Society First World Summit, WSKS 2008, Athens,

Greece, September 24-26, 2008. *Lecture Notes in Computer Science*, LNCS/LNAI 5288 (pp. 573–582), Springer Verlag.

INDEX

A

accountability, 21, 23
activism, 1, 23, 60
advertisements, 18
advocacy, 22
analogical thinking, 16
arrests, 56
association matrixes, 2
attribution, 23
Austria, 55
authenticity, 25

B

bias, 19
black market, 19
blogger, 11, 12, 13, 18, 21, 24, 25, 28, 30, 43
blogosphere, vii, 2, 17, 28, 29, 30, 49, 54, 56, 58, 63
bridges, 25
Broadcasting, 61
browser, 7
buildings, 58

C

campaigns, 6, 26, 27
candidates, 8, 10, 26, 27
case study, vii, 1, 56
CBS, 25
C-C, 59
chimera, 57
China, 20, 61
citizenship, 63
City, 63
clarity, 21
class, 16
classroom, 16
classroom management, 16
cluster analysis, 35
clusters, 2, 35, 36, 41, 45, 46, 47, 48, 49, 50, 51, 53
CNN, 9, 25
coding, 12
coherence, 49, 50, 53
Communist Party, 33, 47
community, 7, 12, 13, 16, 18, 21, 29, 30, 39, 43, 58, 62
compassion, 22
competitive advantage, 5
computer software, 9
conference, 3, 56
confidentiality, 23
configuration, 32
connectivity, vii, 1, 34, 44, 49, 50, 53
connectivity patterns, vii, 1, 34, 44
contradiction, 28
copyright, iv, 7

correlation, 46
cost, 12
covering, 61
creativity, 15
criticism, 25
critics, 57
culture, 15, 18, 20
currency, 16

D

data set, 42
database, 29
democracy, 6, 47
dependent variable, 35
detention, 25
deviation, 37, 39
digital content, 6
directives, 15
discomfort, 22
discrimination, 35
divergence, 48, 49

E

economy, vii, 2, 41, 42, 60
Egypt, 25, 56
election, 24, 26
equality, 23
ethical standards, 21
ethics, 21, 23
European Commission, 4, 62
experiences, 10, 16
exploration, 1, 58
exposure, 16

F

Facebook, 7, 8, 27, 64
fairness, 23
faith, 19
feedback, 14, 16
feelings, 13, 23
filters, 13

focal point blogs, vii, 2, 30, 34, 36, 41, 43
formula, 14, 60
free world, 9
freedom, 3, 25
funding, 23, 26

G

general election, 26, 59
genre, 59
Germany, 25
gill, 58
google, 9
Google, 9, 11
GPC, 5
graph, 31, 34, 40, 46, 51
grassroots, 6
Greece, 18, 28, 33, 44, 47, 64

H

Hawaii, 59
higher education, 64
histogram, 44
honesty, 23
host, 8, 9
household income, 18
hub, 30
Hurricane Katrina, 6
hybrid, 56
hyperlinking, 34
hypertext, 32
hypothesis, vii, 1, 36, 44, 53

I

icon, 6, 7
image, 57
increased access, 16
information exchange, 29
intellectual property, 23
intellectual property rights, 23
intelligence, 3, 63
interface, 5, 6

iPod, 10
Iran, 25
Iraq, 25

J

Japan, 13
journalism, 7, 12, 18, 19, 20, 23, 55, 61, 62, 63, 64
journalists, 7, 18, 19, 20, 21, 63

L

learning, 10, 16, 55, 59, 64
librarians, 56
line graph, 40
LinkedIn, 27
loyalty, 15

M

majority, vii, 13, 18, 20, 30, 36, 37, 44
management, 12, 59
MANOVA, 35
mapping, 32
marketing, 25, 63, 64
mass media, 30
matrix, 31, 34, 35, 39, 40, 42, 49
media, 1, 6, 7, 10, 12, 19, 24, 25, 26, 27, 28, 30, 56, 60, 63
median, 30
Mediterranean, 64
messages, 6, 8, 24
methodology, 31
Microsoft, 7, 14, 15
military, 25
mobile phone, 13
motivation, 59
multivariate statistics, vii
music, 10, 13
MySpace, 7, 10, 27

N

national security, 25
Netherlands, 61
networking, 16, 46
New Zealand, 26
nodes, 31, 32, 34, 39, 40
nursing, 61

O

online learning, 58
online media, 27
openness, 3
opportunities, 24, 39, 63
optimization, 9

P

Parliament, 1
participatory democracy, 5
permit, 35
Perth, 55
photographs, 8, 9, 22
platform, 3, 6
polarization, 2, 36, 48
policy makers, 27
political blogs, vii, 1, 24, 26, 27, 30, 33, 34, 36, 42, 53
political participation, 24
political parties, 59
political party, 48
politics, 5, 6, 7, 10, 18, 23, 24, 26, 27, 30, 57, 58, 61
pop-up windows, 15
presidential campaigns, 26
profit, 14
project, 6, 15
public life, 8, 62
public opinion, 19, 20, 28
public policy, 5, 56

Q

qualifications, 8

R

race, 8, 10
radar, 59, 62
readership, 29, 30
reading, 19, 20
reality, 25
recognition, 19
reconstruction, 56
reliability, 15, 21
reporters, 20, 27
resources, 26, 58
restructuring, 6
rhetoric, 33
rights, iv, 4, 25

S

scaling, 36
scatter, 44
scholarship, 57
self-empowerment, 12
self-expression, 12
self-organization, 1, 54
self-presentation, 63
Senate, 10, 26
sensitivity, 22
sex, 22
shape, 23, 26
Singapore, 60, 61
skewness, 36, 37, 44
SNS, 7, 8
social change, 19
social influence, 59
social learning, 16
social life, 31
social network, vii, 8, 17, 27, 28, 32, 34, 39, 40, 53, 57, 64
Social Networking, vii, 1, 7, 31, 33, 34, 37, 64

social phenomena, 31
social relations, 31
social software, 59
social structure, 31
software, 5, 9, 10, 29, 57
South Dakota, 26
South Korea, 30, 62
speculation, 47
speech, 24, 25
Spring, 60
standard deviation, 37, 39
state control, 25
statistics, 37
survey, 20, 27, 56, 57
synthesis, 17

T

tags, vii, 1, 5, 9, 33, 47
taxonomy, 8
territory, 53
test statistic, 35
thoughts, 11, 13, 16, 18
tourism, 59, 63
training, 19
transformations, 63
Trent Lott, 24
trial, 10
Twitter, 27

U

updating, 12

V

vein, 30, 31
victims, 22
video, 8, 9, 10, 17
videos, 10, 17
vision, 15
vocabulary, 8
voicing, 24
voters, 8, 10, 12

voting, 25

W

weakness, 19
web, 1, 3, 4, 5, 6, 7, 9, 10, 11, 12, 13, 17, 18, 32, 57, 59, 60, 62, 63
Web 2.0, 3, 4, 5, 7, 55, 56, 57, 58, 61, 62
web browser, 5, 6, 17
web pages, 11, 32
weblog, 9, 11, 17, 25, 26, 28, 29, 30, 58, 62
webpages, 9
websites, 3, 5, 6, 10, 11, 17, 31
White House, 10
workplace, 61
World Wide Web, 9
WWW, 55, 59, 62

Y

Yahoo, 9
young adults, 8